ADSORPTIVE REMOVAL OF HEAV GROUNDWATER BY IRON OXIDE BA

T0231321

Adsorptive Removal of Heavy Metals from Groundwater

by Iron Oxide Based Adsorbents

DISSERTATION

Submitted in fulfillment of the requirements of
the Board for Doctorates of Delft University of Technology
and of the Academic Board of the UNESCO-IHE
Institute for Water Education
for the Degree of DOCTOR
to be defended in public on
Friday, 6 December 2013, at 10:00 o'clock
in Delft, The Netherlands

by

Valentine UWAMARIYA

Master of Science, University of the Witwatersrand,
Johannesburg, South Africa
born in Shangi-Nyamasheke, Rwanda.

This dissertation has been approved by the supervisors:

Prof. dr. G.L. Amy
Prof. dr. ir. P.N.L. Lens

Composition of Doctoral Committee:

Chairman	Rector Magnificus Delft University of Technology
Vice-Chairman	Rector, UNESCO-IHE
Prof. dr. G.L. Amy	UNESCO-IHE, Delft University of Technology, supervisor
Prof. dr. ir. P.N.L. Lens,	UNESCO-IHE, Wageningen University, supervisor
Prof. W. van der Meer,	Delft University of Technology
Prof. dr. ir M. Jekel,	University of Technology Berlin
Prof. dr. M.D. Kennedy	UNESCO-IHE, Delft University of Technology
Dr. ir. B. Petrusevski	UNESCO-IHE
Prof. dr. ir. L.C. Rietveld,	Delft University of Technology, reserve member

CRC Press/Balkema is an imprint of the Taylor & Francis Group, an informa business

Published by:
CRC Press/Balkema
PO Box 11320, 2301 EH Leiden, The Netherlands
e-mail: Pub.NL@taylorandfrancis.com
www.crcpress.com – www.taylorandfrancis.com

ISBN 978-1-138-02064-1 (Taylor & Francis Group)

Dedication

This thesis is dedicated to my late father MUNYANGEYO Cléophas.
May his soul rest in peace!

Table of Contents

Foreword

This work would not have been possible without the help of so many people in so many ways. The author extremely thanks the Netherland Government for providing financial assistance through the Netherlands Fellowship Program. I am also grateful to the National University of Rwanda for allowing me the leave to do this research.

I would like to express my deepest gratitude to my promoter Professor Gary Amy together with my co-promoter Professor Piet N.L Lens and my mentor Associate professor Branislav Petrusevski. Without their guidance, valuable discussions, comments and persistence help, this thesis would not have been possible.

I acknowledge with thanks three Msc students who contributed to this work namely, Mr Nikola Stanic, Mr Dibyo Saputro and Mr Muhammed Ahmed Abdullah Bakhamis. I would like to extend my gratitude to UNESCO-IHE laboratory staff namely Mr Fred Kruis, Mr Frank Wiegman, Mr Peter Heerings, Ms Lyzette Robbemont, Mr Ferdi Battes and Mr Don van Galen for the great job for me while I was in Delft. Many thanks are also addressed to Dr Yness M. Slokar for her guidance while I was in the laboratory.

My sincere thanks are addressed to my colleagues Ms. Ingabire Dominique, Dr Sekomo Birame Christian, Dr Babu Mohamed, Dr Kittiwet Kuntiyawichai, Dr Maxmillian Kigobe, Mr Salifu Abdulai, Ms Assiyeh Tabatabai, Mr Chol Abel and Mr Loreen Villacorte.

Special thanks to my best friends Mr Guy Beaujot and Mrs Murungi Caroline and her family for their love and support during my stay in Delft.

Special thanks to my beloved family: my lovely husband Jean de Dieu and sons Elvis and Agnel for their love, prayer support and patience. Thanks for tolerating my absence from home for many months and my long stay for many hours in the office.

To my mother Domina for her comfort and guidance, my brothers Claver and Jean Noël who took care of my sons when I was away and my sister Chantal who was always there when I was in need during my study. All my family members are highly appreciated for their love and support.

Above all, I owe the greatest thanks to the Almighty God for his providence, love and protection.

Abstract

In general groundwater is preferred as a source of drinking water because of its convenient availability and its constant and good quality. However this source is vulnerable to contamination by several substances. Substances that can pollute groundwater are divided into substances that occur naturally and substances produced or introduced by human activities. Naturally-occurring substances causing pollution of groundwater include for example, iron, manganese, ammonium, fluoride, methane arsenic, and radionuclides. Substances resulting from human activities include, for example, nitrates, pesticides, synthetic organic chemicals and hydrocarbons, heavy metals etc.

Acceptable quality limits relative to micropollutant contents in drinking water are becoming increasingly lower and efficient elimination treatment processes are being implemented in order to meet these requirements. Metals contaminants at low concentration are difficult to remove from water. Chemical precipitation and other methods become inefficient when contaminants are present in trace concentrations. The process of adsorption is one of the few alternatives available for such situations. Recent studies have shown that sand and other filter media coated with iron, aluminium, or manganese oxide, hydroxide or oxihydroxide were very good, inexpensive adsorbents which, in some cases, are more effective than the methods usually employed, such as precipitation-coprecipitation or adsorption on granular activated carbon. Selective adsorption can also retain elements that conventional treatment methods are unable to eliminate. This phenomenon was demonstrated after having observed that iron and manganese in particular were more effectively eliminated using old filters than filters containing new sand. This can be explained, in most cases, by a catalytic action of the oxide deposits on the surface of the sand grains.

In this study the adsorption method was used to remove selected heavy metals present as cations (Cd^{2+}, Cu^{2+} and Pb^{2+}) or oxyanions (Cr(VI) and As(V)) using iron oxide coated sand (IOCS) and granular ferric hydroxide (GFH). The effects of pH, natural organic matter (fulvic acid (FA)) and interfering ions (PO_4^{3-}, Ca^{2+}) on the adsorption efficiency were also assessed. The surface complexation modelling for Cd^{2+}, Cu^{2+} and Pb^{2+} adsorption in order to describe the sorption reactions that take place at the surface of the adsorbent was also studied. Batch adsorption tests and rapid small scale column tests (RSST) were used as laboratory methods.

Rwanda uses manly surface water as drinking water and groundwater remains unexplored field and very limited information is available on the quality of this source of drinking water. In this study, groundwater quality was screened in the Eastern province (Nyagatare District), where groundwater is the main source of drinking. For the determination of physico-chemical characteristics of Nyagatare groundwater, 22 parameters were analysed. The results showed that the turbidity and conductivity for all sampled sites are within the range of acceptable values for drinking water. Among the 20 sampled sites, 12 sites have pH values respecting the norms of drinking water, 6 sites have acidic water and 2 sites have alkaline water. For all sampled sites, the dissolved oxygen was found to be low, indicating that Nyagatare groundwater is anoxic. The turbidity is low except for one site, and only four sites are within the acceptable ranges of total alkalinity. Total hardness exceeds the limits for 5 sites and the concentration of major cations (Ca^{2+}, Na^+, K^+ and Mg^{2+}) and major anions (F^-, Cl^-, PO_4^{3-} and SO_4^{2-}) respect the norms of drinking water for all sampled sites. For all sampled sites ammonia concentration is less than 3mg/l except for two sites. NO_2^- and NO_3^- concentrations also respect the WHO (2011) guideline values (2 mg/l and 50 mg/l, respectively). Regarding the concentration of heavy metals, all sampled sites have values of Fe^{2+} exceeding the value of 0.3 mg/l that is the upper acceptable concentration in most national drinking water standards (including Rwanda) and ten sites have values of Mn^{2+} exceeding the value of 0.1 mg/l that is recommended by several national standards to avoid esthetic and operational problems. For other heavy metals, Zn^{2+} respects the norm for all sampled sites except for all Rwempasha and Rwimiyaga sites. Even if the main focus of this research is the removal of heavy metals, the concentrations of Pb^{2+}, Cd^{2+}, Cu^{2+}, As and Cr in Nyagatare groundwater were found to be below the detection limits.

With a Piper diagram representation, most of sampled sites are found to be mainly sodium and potassium type and, for few of them, no dominant type of water could be found. In terms of anions, few sites have chloride groundwater type, one has bicarbonate groundwater type and others have no dominant anions. The total hardness varied between 10 and 662 mg/l, 7 samples fall under soft class, 3 samples fall under moderately hard class, 7 samples fall under hard and 3 samples fall under very hard class. The calculation of percentage of Na^+, the residual sodium carbonate (RSC) and sodium adsorption ratio (SAR) showed that Nyagatare groundwater is suitable for irrigation. The Nyagatare district has abundant granite and granite igneous rocksrocks, and this can explain the source of fluoride found in groundwater. The source of EC, TDS, ammonia and nitrite in Nyagatare groundwater can be related to human activities e.g. application of fertilizers and manure.

Principal component analysis (PCA) results showed that the extracted components represent the variables well. The extracted six components explain nearly 94% of the variability in the original 22 variables, so that one can considerably reduce the complexity of the data set by using these components, with only 6% loss of information. The first component was most highly correlated with fluoride, pH and sulfate; the second component was most strongly correlated with calcium and total hardness, while the third component is most strongly correlated with total alkalinity. The fourth the fifth and the sixth components are mostly correlated with potassium, iron and magnesium, respectively.

The effects of calcium on the equilibrium adsorption capacity of As(III) and As(V) onto iron oxide coated sand and granular ferric hydroxide were investigated through batch experiments, rapid small scale column tests and kinetics modeling. Batch experiments showed that at calcium concentrations ≤ 20 mg/l, high As(III) and As(V) removal efficiencies by IOCS and GFH were observed at pH 6. An increase of the calcium concentration to 40 and 80 mg/l reversed this trend giving higher removal efficiency at higher pH (8). The adsorption capacities of IOCS and GFH at an equilibrium arsenic concentration of 10µg/l were found to be between 2.0 and 3.1 mg/g for synthetic water without calcium and between 2.8 and 5.3 mg/g when 80 mg/l of calcium was present at all studied pH values. After 10 hours of filter run in rapid small scale column tests, and for approximately 1000 Empty Bed Volumes filtered, the ratios of C/C_o for As(V) were 26% and 18% for calcium-free model water; and only 1% and 0.2% after addition of 80 mg/l of Ca for filter columns with IOCS and GFH, respectively. The adsorption of As(III) and As(V) onto GFH follows a second order reaction irrespective of calcium presence in model water, while the adsorption of As(III) and As(V) onto IOCS follows a first-order reaction in calcium-free model water, and moves to the second reaction order kinetics when calcium is present. Based on the intraparticle diffusion model, the main controlling mechanism for As(III) adsorption is intraparticle diffusion, while the surface diffusion contributes greatly to the adsorption of As(V).

The effect of PO_4^{3-} on the adsorptive removal of Cr(VI) and Cd^{2+} was assessed using IOCS and GFH as adsorbents. Batch adsorption experiments and RSSCT were conducted using Cr(VI) and Cd^{2+} containing model water at pH 6, 7 and 8.5. The best Cr(VI) and Cd^{2+} adsorption was observed at pH 6. GFH showed much better removal of Cr(VI) than IOCS, while IOCS removed Cd^{2+} better than GFH. Increasing PO_4^{3-} concentrations in the model water from 0 to 2 mg/L, at pH 6, induced a strong decrease in Cr(VI) removal efficiency from 93% to 24% with GFH, and from 24% to 17% with IOCS. A similar trend was observed at pH 7 and 8.5. An exception was for Cr(VI) removal with IOCS at pH 8.5, which was not affected by the PO_4^{3-} addition. Cd^{2+} was well removed by both GFH and IOCS, contrary to Cr(VI) which was better removed by IOCS. The effect of PO_4^{3-} is clearly seen at pH 6 when there is no precipitation of Cd^{2+} in the solution. At pH 8.5, the precipitation is the main removal process, as it represents around 70% removal of Cd^{2+}. The isotherm constants K for different combinations confirms the inhibition of Cr(VI) and enhancement of Cd^{2+} adsorption with addition of PO_4^{3-}. The same conclusion is confirmed by the results from rapid small scale column tests. The mechanism of Cr(VI) adsorption by GFH and IOCS is likely a combination of electrostatic attraction and ligand exchange while the mechanism of Cd^{2+} removal at lower pH of 6 was sufficiently energetic to overcome some electrostatic repulsion.

The effects of pH and Ca^{2+} on the adsorptive removal of Cu^{2+} and Cd^{2+} was also assessed in batch adsorption experiments and kinetics modelling. It was observed that Cu^{2+} and Cd^{2+} were not stable at pH values considered (6, 7 and 8), and the precipitation was predominant at higher pH values, especially for Cu^{2+}. The increase in Ca^{2+} concentration also increased the precipitation of Cu^{2+} and Cd^{2+}. It was also observed that Ca^{2+} competes with Cu^{2+} and Cd^{2+} for surface sites of the adsorbent. The presence of calcium diminishes the number of available adsorption sites of IOCS and GHF, resulting in lower removal of cadmium and copper. Freundlich isotherms for cadmium removal by IOCS showed that the adsorption capacity of IOCS decreased when calcium was added to the model water. The kinetics modelling revealed that the adsorption of Cd^{2+} onto IOCS is likely a second-order reaction.

The effects of fulvic acid on adsorptive removal of Cr(VI) and As(V) was also assessed. Batch adsorption experiments and characterization of IOCS and GFH by SEM/EDS were performed at different pH levels (6, 7 and 8). The surface of the virgin IOCS showed that Fe and O represent about 60 to 75% of the atomic composition, while carbon concentration was about 10%. The surface analysis of GFH showed that Fe and O represent, about 32% and 28% of the chemical composition, respectively. The adsorption experiments with simultaneous presence of As(V) and FA on the one hand, and Cr(VI) with FA on the other hand, revealed that the role of FA was insignificant at all almost pH values for both IOCS and GFH. Some interference of FA on the removal of As(V) and by IOCS or GFH was only observed at pH 6. It was also found out that organic matter (OM) was leaching out from the IOCS during experiments. The use of EEM revealed that humic-like, fulvic-like and protein-like organic matter fractions are present in the IOCS structure.

Removal of selected heavy metals, namely Cd^{2+}, Cu^{2+}, and Pb^{2+}, by IOCS was also screened in a series of batch adsorption experiments conducted at different pH. Studies metals were present in model water as single or together with some other metals. Results from adsorption experiments using model water with a single metal, and using IOCS as an adsorbent, showed that all metals included in the study can be very effectively removed with total removal efficiency as over 90% at all pH levels studied. XRF analysis showed that IOCS contains mainly hematite (Fe_2O_3) (approximately 85% of the total mass of minerals that could be identified by XRF). Chemical analysis revealed that the main constituent of IOCS is iron, representing 32% on mass basis. Potentiometric mass titration (PMT) gave a value of pH of zero point charge of 7.0. The percentage of metals removed through precipitation was found to be metal specific: the highest for Cu (25%) and the lowest for Cd (2%) at pH 8. Concurrent presence of competing metals did not have a pronounced effect on the total metal removal efficiency, with the observed reduction of total removal efficiency of Cu, Cd and Pb between 1 and 4%. In terms of adsorption capacity, a competitive effect of metals was not observed except for Pb and Cu at pH 8 where the adsorption was decreased for 13% and 22%, respectively.

Complexation modelling showed two type of complexes, one type associated with a weak site (Hfo_wOCd$^+$, Hfo_wOCu$^+$, Hfo_wOPb$^+$), and the other associated with a strong site (Hfo_sOCd$^+$, Hfo_sOCu$^+$, Hfo_sOPb$^+$), formed for all metals studied. Precipitation of Pb and Cu observed in batch experiments was confirmed in modelling at pH $\geq$ 6.75. IOCS, being an inexpensive and easily available adsorbent, can be used to treat water contaminated with heavy metals like Cd, Cu and Pb. However, pH is an important factor to be considered if one has to avoid precipitated metals which will finish in liquid waste (backwash water) , especially for the removal of Cu and Pb.

Chapter 1: Introduction

Water is generally obtained from two principal natural sources: surface water such as fresh water lakes, rivers, streams and groundwater such as borehole water (McMurry and Fay, 2004; Mendie, 2005). Water is one of the essentials that supports all forms of plant and animal life (Vanloon and Duffy, 2005) and it has unique chemical properties due to its polarity and hydrogen bonds which means it is able to dissolve, absorb, adsorb or suspend many different compounds (WHO, 2007). Groundwater is the major source of drinking water in the world because of its availability and constant quality. Groundwater is also the preferred source of drinking water in rural areas, particularly in developing countries, because no treatment is often required and the water source is often located near consumers. However, in nature, water is not pure as it acquires contaminants from its surroundings and those arising from humans and animals as well as other biological activities (Mendie, 2005). This chapter reviews the literature on the quality, the problems related to groundwater pollution, and the different techniques used in the analysis and removal of heavy metals from groundwater. This chapter also provides the scope of the thesis.

1.1 Overview of groundwater availability and quality

Groundwater is water below the land surface that fills the spaces between the grains or cracks and crevices in rocks. It is derived from rain and percolation down through the soil. Groundwater has a number of essential advantages when compared with surface water: it is of higher quality, better protected from possible pollution, less subject to seasonal and perennial fluctuations, and much more uniformly spread over large regions than surface water. Groundwater also can be available in places where there is no surface water. Putting groundwater well fields into operation is also less costly in comparison to what is needed for surface water which often requires considerable capital investments. These advantages coupled with reduced groundwater vulnerability to pollution particularly have resulted in wide spread groundwater use for water supply. Currently, 97% of the planet's liquid freshwater is stored in aquifers. Many countries in the world consequently rely to a large extent on groundwater as a source of drinking water. Table 1.1 shows that 2 billion people rely on groundwater as the only source of drinking water (Sampat, 2000).

Table 1.1: Groundwater use for drinking water production by region

Region	Share of drinking water from groundwater (%)	People served (million)
Asia and Pacific	32	1000-2000
Europe	75	200-500
Latin America	29	150
United states	51	135
Australia	15	3
Africa	?	?
World		2000 (2.0 billion)

(Source: Sampat, 2000)

Groundwater is the major source of drinking water in many countries all over the world (Table 1.2). Unfortunately little is known about use of groundwater for drinking water in Africa. Table 2 shows that groundwater is extensively used as an important source of public water supply in Europe, especially in Denmark where groundwater represents 100% of the drinking water production. In rural areas of the United States and India, groundwater also represents the primary source of potable water (96 % and 80 % respectively). Groundwater meets over 75% of the water needs of Estonia, Iceland, Russian Federation, Jamaica, Georgia, Swaziland, Mongolia, Libya and Lithuania (Vrba, 2004). Groundwater in Tunisia represents 95% of the country's total water resources, in Belgium it is 83%, in the Netherlands, Germany and Morocco it is 75%. In most European countries (Austria, Belgium, Denmark, Hungary, Romania and Switzerland) groundwater use exceeds 70% of the total water consumption (Vrba, 2004). In many nations, more than half of the withdrawn groundwater is for domestic water supplies and globally it provides 25 to 40 % of the world's drinking water (UNEP 2003).

Table 1.2: Groundwater use for drinking water production by selected countries

Region	Country	Percentage (%)
Europe	Denmark	100
	Germany	75
	Slovakia	82
	The Netherlands	75
	Belgium	83
	The United Kingdom	27
Asia	India (rural)	80
	Philippines	60
	Thailand	50
	Nepal	60
Africa	Ghana	45
	Morocco	75
	Tunisia	95
America	United States (rural)	96

(Source: Vrba, 2004)

Various human activities can result in significant changes in the conditions of the groundwater resources formation, causing depletion and pollution. Groundwater pollution in most cases is a direct result of environmental pollution. Groundwater is polluted mainly by nitrogen compounds (nitrate, ammonia and ammonium), petroleum products, phenols, iron compounds, and heavy metals (copper, zinc, lead, cadmium, mercury) (Vrba, 2004).

Groundwater is closely interrelated with other components of the environment. Any changes in atmospheric precipitation inevitably cause changes in the groundwater regime, resources and quality. Vice versa, changes in groundwater cause changes in the environment. Thus, intensive groundwater exploitation by concentrated water well systems can result in a decrease in surface water discharge, land surface subsidence, and vegetation suppression due to groundwater withdrawal. Groundwater pumping can extract mineralized groundwater not suitable for drinking in deep aquifers, and can draw in saline seawater in coastal areas. All of these circumstances should be considered when planning groundwater use. Even if groundwater is less vulnerable to human impacts than surface water, once groundwater is polluted, remediation can be relatively long term (years), technically demanding and costly (Vrba, 1985).

Groundwater systems are replenished by precipitation and surface water. Globally groundwater circulation is less than atmospheric and surface waters, but what is stored beneath the earth's surface is orders of magnitude larger. Its total volume represents 96% of all earth's unfrozen fresh water (Shiklomanov and Rodda, 2003).

About 60% of the groundwater withdrawn is used by agriculture in many countries where arid and semi-arid climates prevails; the rest is almost equally divided between the domestic and industrial sectors (UNEP, 2003). Table 1.3 shows the 10 countries that use largely groundwater for agriculture, domestic and industry uses.

Table 1.3: Top 10 countries with largest groundwater extraction (Vrba, 2004)

Country	GW Abstraction (Km³/year)	Reference Year	Main use (main sector)	% of main sector demands covered by GW
India	190	1990	Agriculture	53
USA	115	2000	Agriculture	42
China	97	1997	Agriculture	18
Pakistan	60	1991	Agriculture	34
Iran	57	1993	Agriculture	50
Mexico	25	1995	Agriculture	30
Russia Federation	15	1996	Domestic	80
Saudi Arabia	14	1990	Agriculture	96
Italy	14	1992	Agriculture	
Japan	14	1995	Industry	35

1.2 Heavy metals in groundwater

1.2.1. Introduction

Groundwater contamination is one of the most important environmental issues today and a big problem in many countries (Vodela *et al.*, 1997, Sharma and Busaidi, 2001). Among the wide diversity of contaminants affecting water resources, heavy metals receive particular concern considering their strong toxicity even at low concentrations (Marcovecchio *et al.*, 2007). Most heavy metal ions are non-degradable ions, persistent in the environment and toxic to living organisms. Therefore, the elimination of heavy metal ions from water is important to protect public health.

Momodu and Anyakora (2010) reported that industries such as plating, ceramics, glass, mining and battery manufacturing are considered as the main sources of heavy metals in local water streams, which can cause the contamination of groundwater with heavy metals. In addition, heavy metals which are commonly found in high concentrations in landfill leachate also are a potential source of pollution for groundwater (Aziz et al., 2004, Marcovecchio *et al.*, 2007). The practice of landfill systems as a method of waste disposal is the main method used for solid waste disposal in many developing countries. However, a "landfill" in a developing country's context is usually unprotected and shallow (often not deeper than 50 cm). This is usually far from standard recommendations (Mull, 2005; Adewole, 2009) and contributes to the contamination of air, water (surface and groundwater) and soil and is a risk to human and animals. Besides, it has been shown that even protected landfills can be inadequate in the prevention of groundwater contamination (Lee and Lee, 2005).

Atomic weights of heavy metals range between 63.5 and 200.6. They have a specific gravity of five times higher than water. Heavy metals in water can be in colloidal, particulate and dissolved phases (Adepoju-Bello *et al.*, 2009). Some heavy metals are essential to human life like cobalt, copper, iron, manganese, molybdenon and zinc which are needed at low levels as a catalyst for enzyme activities (Adepoju-Bello *et al.*, 2009). However, excess exposure to heavy metals can result in toxicity.

The quality of groundwater sources is affected by the characteristics of the media through which the water passes on its way to the groundwater zone of saturation (Adeyemi *et al.*, 2007), thus, the heavy metals discharged by industries, traffic, municipal wastes, hazardous waste sites as well as from fertilizers for agricultural purposes and accidental oil spillages from tankers can result in a steady rise in contamination of groundwater (Vodela *et al.*, 1997; Igwilo *et al.*, 2006).

1.2.2 Chemistry and toxicity of heavy metals

Depending on the nature and quantity of the metal ingested, heavy metals can cause serious health problems (Adepoju-Bello and Alabi, 2005). Their toxicity is related to the formation of complexes with proteins, in which carboxylic acid (–COOH), amine (–NH$_2$), and thiol (–SH) groups are involved. When metals bind to these complexes, important enzyme and protein structures are affected. The most dangerous heavy metals that humans are exposed to are aluminium, arsenic, cadmium, lead and mercury. Aluminium has been associated with Alzheimer's and Parkinson's disease, senility and presenile dementia. Arsenic exposure can cause among other illness or cancer, abdominal pain and skin lesions. Cadmium exposure produces kidney damage and hypertension. Lead is a commutative poison and a possible human carcinogen (Bakare-Odunola, 2005), while for mercury, the toxicity results in mental disturbance and impairment of speech, hearing, vision and movement (Hammer and Hammer Jr., 2004). In addition, lead and mercury may cause the development of auto-immunity in which a person's immune system attacks its own cells, which can lead to joint diseases and ailment of the kidneys, circulatory system and neurons. At higher concentrations, lead and mercury can cause irreversible brain damage.

The chemistry, source and toxicity of some heavy metals are detailed below.

1.2.2.1 Arsenic

Arsenic is a naturally occurring element that is tasteless and odorless. Arsenic occurs in an inorganic form in the aquatic environment; resulting from the dissolution of solid phases as As$_2$O$_3$ (arsenolite), As$_2$O$_5$ (arsenic anhydre) and AsS$_2$ (realgar) (Faust and Aly, 1998). Arsenic can exist in several oxidation states including the +5, +3, +1, and -3 valences and rarely in the elemental form. The most common valence states of arsenic in geogenic raw water sources are As(V) or arsenate and As(III) or arsenite (Irgolic, 1982). In the pH range of

4 to 10, the prevalent As(III) species is neutral in charge, while the As(V) species is negatively charged.

As a compound of underground rocks and soil, arsenic works its way into groundwater and enters the food chains through either drinking water or eating plants and cereals that have absorbed the mineral. Daily consumption of water with greater than 0.01 mg/l of arsenic, less than 0.2 % of the fatal dose, can on long teerm lead to problems with the skin as well as circulatory and nervous systems. If arsenic builds up in the human body, open lesions, organ damages (such as deafness), neural disorders and organ cancer, often fatal, can develop (Pal, 2001). The maximum acceptable concentration of arsenic in drinking water recommended by the World Health Organization is 10µg/l (WHO, 2011).

1.2.2.2 Copper

In the aquatic environment, copper can exist in three broad categories: particulates, colloidal and soluble. The dissolved phase may contain both free ions as well as copper complexed to organic and inorganic ligands. Copper forms complexes with hard bases like carbonate, nitrate, sulfate, chloride, ammonia, hydroxide and humic materials. The formation of insoluble malachite ($Cu_2(OH)_2CO_3$)) is a major factor in controlling the level of free copper (II) ions in aqueous solutions. Copper complexes of oxidation state (+1), (+2) and (+3) are known although Cu(+2) is more common. Cu(+1) is a typical soft acid. The Cu(+2) ion is the major species in water at pH up to 6; at pH 6.0-9.3 aqueous $CuCO_3$ is prevalent and at pHs 9.3-10.0, the aqueous $[Cu(CO_3)_2]^{2-}$ ion predominates (Stumm and Morgan, 1996). At a pH below 6.5, free copper ions are the dominant copper containing species. Alkalinity has a profound effect on the free copper ion concentration (Snoeyink and Jenkins, 1980).

Copper is of particular interest because of its toxicity and its widespread presence in the industrial applications like electroplating, metal finishing and paint industries. The presence of copper in drinking water may cause itching, dermatitis, keratinization of the hands and sole of the feet. Therefore the concentration of copper in drinking water must be reduced to levels that satisfy the environmental regulation for various bodies of water (Huang, 1989). Dissolved copper imparts a color and unpleasant, metallic, bitter taste to drinking water. Staining of laundry and plumping fixtures occur when the copper concentration in water exceeds 1.0 mg/l. Vomiting, diarrhea, nausea and some acute symptoms presumably due to local irritation by ingested copper (II) ions have been described in several cases. The maximum acceptable concentration of copper in drinking water based on health considerations is 2.0 mg/l (WHO, 2011).

1.2.2.3 Cadmium

Cadmium is most commonly found associated with zinc (Zn) in carbonate and sulfide ores. It is also obtained as a by-product in the refining of other metals. The aqueous chemistry of cadmium is, for the most part, dominated by Cd^{2+}, $CdCO_3(s)$ (otavite) and $Cd(OH)_2(s)$ (Faust and Aly, 1998). The solubility of cadmium in water is influenced by its acidity. For pH values greater than 10, the solubility of $CdCO_3$ is about 300 µg/l and the solubility of cadmium hydroxide at pH 10 is 44 µg/l for the aged form and 225 µg/l for the fresh precipitate. In the pH range of groundwater the various complexes of cadmium are present but at pH values lower than 6, the above mentioned complexes are not present in significant concentrations (Snoeyink and Jenkins, 1980). The species Cd^{2+}, $CdOH^+$ and $Cd(OH)_2$ contribute increasingly to the solubility of cadmium in the pH range of 4 to 9, as reported by Chin (2002).

Cadmium can be present in groundwater from a wide variety of sources in the environment and from industry. One source is ingestion of grown foodstuffs, especially grain and leafy vegetables, which readily absorb cadmium from the soil. The cadmium may occur in groundwater naturally or as a contaminant from sewage sludge, fertilizers, polluted groundwater or mining effluents (Hu, 1998).

Cadmium has no essential biological function and is extremely toxic to humans. In chronic exposure, it also accumulates in the body, particularly in the kidneys and the liver. Acute poisoning from inhalation of fumes and ingestion of cadmium salts can also occur and death has been reported from self-poisoning with cadmium chloride (Baldwin and Marshall, 1999). An investigation on human health risks shows that cadmium causes many diseases if inhaled at higher doses. The disease called Itai-Itai (Ouch-Ouch) is well documented in Japan. It was named Itai-Itai due to pain caused by the decalcification and final fracturing of bones, which are symptomatic of cadmium poisoning (WHO, 1996). Based on the possible toxicity of cadmium, the WHO health-based guideline value for drinking water is 3μg/l (WHO, 2011).

1.2.2.4 Lead
Lead contamination can occur in groundwater due to mining and smelting activities, battery plant emissions, battery reprocessing plant wastes, automotive exhaust emissions, leaded fuel spills, incinerator ash disposal and municipal or industrial land fill leachates. The behavior of lead in water is a combination of precipitation equilibrium and complexing inorganic and organic ligands. The binding of lead to negatively charged organic surfaces increases rapidly above pH 5, but is decreased somewhat by competition with soluble organic substances and metal chelates (James and Ramamoorhty, 1984). The predominant form of lead in natural waters is a function of the ions present, their concentrations, the pH and the redox potential. The solubility of lead is 10μg/L above pH 8, while near pH 6.5 the solubility can approach or exceed 100μg/l.

Lead is an ubiquitous trace constituent in the environment that has been known for centuries to be an accumulative metabolic poison. The groups most susceptible to lead poisoning are foetuses and children. The effect of lead on the central nervous system can be particularly serious. Developing central nervous systems of children may be affected, leading to hyperactivity, irritability, headaches, and learning and concentration difficulties. The maximum acceptable concentration of lead in drinking water is 0.01 mg/L (WHO, 2011).

1.2.2.5 Chromium
Chromium in (ground) water is found in two valences (Cr(III) and Cr(VI)). Many chromium, especially chromium (III), compounds are relatively water insoluble. Chromium (III) oxide and chromium (III) hydroxide are the only water soluble compounds. Chromium (VI)

compounds are stable under aerobic conditions, but are reduced to chromium (III) compounds under reducing conditions. The reverse process is another possibility in an oxidizing environment. The dominant Cr(III) species occurring in groundwater depend on pH, $Cr(OH)_2^+$ being the dominant species in natural groundwater with a pH between 6 and 8 (Calder, 1988). Cr(VI) in aqueous solution exists almost exclusively in the form of oxy-anions (CrO_4^{2-}, $Cr_2O_7^{2-}$). In dilute solutions (< 1 ppm), the predominant form is CrO_4^{2-}; being negatively charged, it does not complex with anionic particulate matter. However, Cr(VI) anions are adsorbed on positively charged surfaces, such as the oxides and hydroxides of Fe, Mn and Al. Adsorption of Cr(VI) on these adsorbents is usually limited and decreases with increasing pH (Slooff et al. 1990). Hence, Cr(VI) is more mobile than Cr(III). Cr(III) species are predominant at pH values less than 3, and, at pH values above 3.5, hydrolysis of Cr(III) yields trivalent chromium hydroxy species ($Cr(OH)^{2+}$, $Cr(OH)_2^+$, $Cr(OH)_3^0$, $Cr(OH)_4^-$).

Chromium is a dietary requirement for a number of organisms. This, however, only applies to trivalent chromium. Hexavalent chromium is very toxic to flora and fauna. The drinking water guideline concentration for total chromium in drinking water is 50µg/l (WHO, 2011). This guideline value is provisional due to uncertainties related to the health effect and that lower guideline value could be expected in the future. Total chromium has been specified because of difficulties in analyzing the hexavalent form only.

1.3 Technologies to remove heavy metals

1.3.1 Introduction
In order to meet regulatory standards, and in view of their toxicity, heavy metals need to be removed from contaminated water. The conventional processes used in the removal of heavy metals from contaminated water include chemical precipitation, reverse osmosis, adsorption, ion exchange and electrochemical deposition (Meena et al, 2005 and Barakat 2011). However, chemical precipitation requires a large amount of chemicals and produces large amount of sludge with environmental impacts of its disposal (Aziz et al., 2008). The ion exchange also has some disadvantages such as the difficult to handle concentrated metal solution. In addition ion exchange resins get easily fouled by organics and other solids in the treated water (Barakat 2011).

1.3.2 Removal of arsenic
Several technologies have been used to remove arsenic from groundwater. These include coagulation, precipitation, adsorption, ion exchange, membrane filtration, sub-surface (in situ) removal and biological processes. However, not all of the above-mentioned treatment technologies have found useful application in practice for the removal of arsenic from ground water. Out of the existing methods of arsenic removal, some of the methods like coagulation, oxidation followed by ion–exchange, and adsorption are well established, while others like adsorption by metal oxide coated sand, and electrolytic reduction of As (V) to As (III) and subsequent precipitation are still being investigated. Adsorption of As(V) and As(III) on commercial adsorbents (Pal 2005) and IOCS (Petrusevski *et al.* 2001) showed promising results. The removal efficiency for As(III) is poor compared to that for As(V) by any of the conventional technologies for elimination of arsenic from water. For effective removal of

arsenic from water, with conventional technologies (e.g. coagulation, RO, IEX) a complete oxidation of As(III) to As(V) is required (Karaschunke and Jekel, 1989*).*

1.3.3 Removal of copper

Tekeste (2003) also conducted batch experiments on the removal of Cd, Cu, Cr, Ni and Pb by IOCS. He found that IOCS exhibited very high adsorption capacities for all heavy metals studied. According to the study conducted by Devendra (2007) on the adsorptive removal of heavy metals from urban storm water run-off, it has been shown that IOCS and GFH are good adsorbents for the removal of copper. Removal mechanism of copper with GFH could not be explained on the basis of electrostatic attraction under the experimental conditions applied.

Barakat (2005) reported that Cu(II) and Zn(II) were adsorbed at neutral and alkaline pH by zeolite. Feng et al. (2004) investigated Cu(II) and Pb(II) removal using iron slag. A pH range from 3.5 to 8.5 for Cu(II) was found to be optimal. Alinnor (2007) used fly ash from coal-burning for removal of Cu(II) and Pb(II) ions. Barakat (2005) reported that, the adsorbed Cu(II) aqueous species on zeolite can undergo surface hydrolysis reaction as pH rises. This yields a series of surface Cu(II) complexes such as $TiO-CuOH^+$, $TiO-Cu(OH)2$, and $TiO-Cu(OH)3$ species. TiO_2 particles are negatively charged at pH 6, and so complete Cu(II) adsorption was achieved at such pH range

1.3.4. Removal of lead

Alum and ferric sulfate coagulants were effective for lead removal in jar tests and pilot runs over the wide pH range 6-10 (Sorg, 1986). Lime softening was also found to be effective in the removal of lead. Sorg (1986) speculated on the use of a strong acid cation exchange process for Pb removal from groundwater. The selectivity series for cations indicates that this process of strong cation exchange should be effective. The order of the removal of cation is as follow:

$Ba^{2+} > Pb^{2+} > Sr^{2+} > Ca^{2+} > Ni^{2+} > Cd^{2+} > Cu^{2+} > Co^{2+} > Zn^{2+} > Mg^{2+}$ and $Ag^+ > Cs^+ > Rib^+ > K^+ > NH_4^+ > Na^+ > H^+ > Li^+$

Few studies have been conducted on the removal of Pb by powered activate carbon (PAC) or granular activated carbon (GAC). Based on the results from laboratory jar tests conducted at pH 7.3, it was found that 10 mg/l of PAC effected 98% lead reduction of (Faust and Aly, 1998). In a parallel study using GAC in a pilot plant, Pb removal from model water with 0.11 to 0.20 mg/l of Pb was higher than 95%.

1.3.5 Removal of chromium

The removal of Cr(III) at neutral pH by NaA zeolite was studied and found to be effective (Basaldella et al.2007). Barakat (2008a) used 4A zeolite which was synthesized by dehydroxylation of low grade kaolin and reported that Cr(VI) was adsorbed at acidic pH. The adsorption of Cr(VI) on sawdust treated with 1,5-disodium hydrogen phosphate was also tested at pH 2 by Uysal and Ar (2007) . Babel and Kurniawan (2004) investigated the coconut shell charcoal (CSC) modified with oxidizing agents and/or chitosan for Cr(VI) removal. The maximum Cr(VI) removal by rice husk took place at pH 2.0 as reported by

Bishnoi et al. (2003). Cr(VI) removal was also tested on rice hull, containing cellulose, lignin, carbohydrate and silica (Tang et al., 2003). The maximum Cr(VI) adsorption capacity of 23.4 mg/g was reported to take place at pH 2.

The coagulation process by ferric sulfate and alum was found to be ineffective for Cr(VI). However when ferrous sulfate was employed, nearly 100% of Cr(VI) was removed. Cr(VI) was reduced to Cr(III) by the Fe(II) ion, with subsequent precipitation as Cr(OH)$_3$. Several studies of the removal of Cr by PAC and GAC have been conducted. For example, GAC has been used in the removal of Cr(VI) over the wide pH range from 2 to 10. The highest removal occurred at pH 2; there was an exponential decrease of effectiveness with pH increase to about 6. At this pH there was virtually no chromium reduction (Faust and Aly, 1998). Recently, studies on the removal of Cr by IOCS gave promising results (Tessema, 2004).

1.3.6 Removal of cadmium

The removal of Cd^{2+} has been extensively studied. Potential of solid waste from sugar industry (bagasse fly ash) to remove Cd^{2+} from synthetic solution in the pH range varying between 6.0 and 6.5 was tested by Gupta et al. (2003). The use of chitosan derivatives containing crown gave high adsorption capacity for Cd^{2+} removal Yi et al.(2003). ZnAl$_2$O$_4$– TiO$_2$ UF membranes were successfully used by Saffaj et al. (2004) to remove Cd^{2+} ions from synthetic solution, with reported 93% removal achieved. The electrodialytic removal of Cd^{2+} from wastewater sludge was studied by Jakobsen et al., (2004) at the liquid/solid (mL/g fresh sludge) ratio varying between 1.4 and 2. The Cd^{2+} removal in the three conducted experiments was between 67 and 70%. According to Devendra (2007), the removal of cadmium with IOCS observed in batch adsorption experiments was likely due to electrostatic attraction between the positively charged species of cadmium and negatively charged surface of IOCS. However, there no removal of cadmium with GFH was observed in conducted rapid small scale column tests (RSSCT).

1.4 Properties of Iron oxides

This study aims at using iron oxide based adsorbents in removing heavy metals from contaminated groundwater. This item gives an overview of the properties of different iron oxides adsorbents.

Iron is a very good element for decontamination of the environment (Benjamin, 1983) since it is both a reductant and oxidant, non-toxic and not very expensive. Furthermore, like other oxides and hydroxides, it is an exchanger of either cations or anions and, in some cases; it can even fulfill both roles at the same time (Vaishya et al., 2003). Sands coated with metal oxides can retain either cationic or anionic forms of metals (Lo et al., 1997). Iron is the fourth most abundant element in the earth's crust and the second most abundant metal (Jambor and Dutrizac, 1998). It occurs as Fe(II) and Fe(III) in a diversity of minerals including many types of iron oxides. The iron oxides are in fact oxides, hydroxides or oxihydroxides. Most of the compounds are thermodynamically stable in natural systems (e.g. goethite, hematite, and magnetite) while others can be designated as intermediates only (ferrihydrite, maghemite)

(Jambor and Dutrizac, 1998). The different iron oxides have their own characteristics and qualities, governed by the mineral structure. In the presence of water, the surfaces of these oxides are generally covered with surface hydroxyl groups. These functional groups contain the same donor atoms as found in functional groups of soluble ligands (S-OH). Deprotonated groups (S-O-) behave as Lewis bases and the sorption of metal ions (and protons) can be understood as competitive complex formation. The adsorption of ligands (anions and weak acids) on metal oxide surfaces can also be compared with complex formation reactions in solution (Stumm, 1992).

1.4.1 Ferrihydrite

Ferrihydrite is found in many systems. It occurs in nature in waters and sediments, in soils, in mine waste and acid mine drainage, and even in meteorites (Jambor and Dutrizac, 1998). Ferrihydrite forms through rapid precipitation and oxidation of aqueous Fe(II), particularly in complex and contaminated systems. Ferrihydrite is thermodynamically meta-stable and will over time transform into more stable members of the group, typically hematite and goethite, unless stabilized in some way. Ferrihydrite has an extensive specific surface area and a high adsorptive capacity and, can, thus, retain large amounts of foreign ions by sorption. Ions may, in addition, be retained by sorption to "inner surfaces" in the aggregated particles. Substitution of foreign cations in the ferrihydrite structure has never been proven even though a great number of metal containing co-precipitates have been studied (Jambor and Dutrizac, 1998).

1.4.2 Goethite

Goethite is frequently found in nature. It is one of the thermodynamically most stable iron oxides at ambient temperatures, and it is, for this reason, often found as an end product of many transformations. Goethite can form directly in solution via a nucleation-crystal growth process, e.g. from ferrihydrite via dissolution and re-precipitation. It is generally formed in competition with hematite in aqueous systems. The resultant ratio between the goethite and the hematite depends on factors like pH, temperature and ionic strength of the solution. Foreign cations like Al, Cd, Co, Cr, Cu, Mn, Ni and Zn can readily replace Fe in the goethite structure by substitution (Gerth, 1990).

1.4.3 Hematite

The red hematite is often found in nature, particularly in tropical or subtropical soils. Like goethite, hematite is extremely stable and, hence, often the end member of transformations of other iron oxides. Hematite can form in several ways but one common way implies an internal rearrangement and dehydration of ferrihydrite (Schwertmann and Cornell, 1991). In aqueous systems, hematite generally forms in competition with goethite. Also a thermal dehydration can generate a phase of hematite, depending on the redox conditions. Foreign cations like Al, Cu, Cr, and Mn can substitute for Fe in hematite (Singh and Gilkes, 1992).

1.4.4 Maghemite

Maghemite is invariably found in soils, primarily in the tropics and the subtropics. As implied by its name, maghemite is a combination of magnetite and hematite. It has the structure of magnetite and the composition of hematite and it can, consequently, be considered a fully oxidized magnetite. Maghemite can form by heating magnetite in an oxidizing atmosphere. This formation of maghemite from magnetite requires the ejection of 11 % Fe, and it holds, therefore, vacancies in the structure. Maghemite integrates foreign cations like Al, Co, Cu, Cr, Mn, Ni, and Zn (Sorensen, 2001).

1.4.5 Magnetite

The presence of magnetite in nature is often a result of biological processes, but it can also have a lithogenic origin (Schwertmann and Cornell, 1991). Magnetite has the structure of an inverse spinel and it differs from most other iron oxides in that it contains both Fe(II) and Fe(III). Magnetite can form directly by oxidative hydrolysis of a Fe(II) solution, and heating of Fe(III) compounds under strongly reducing conditions.

Table 1.4: Properties of some iron oxides (Cornell and Schwertmann, 1991)

Structure	Formula	Surface area (m^2/g)	Substituting cations
Ferrihydrite	$Fe_5HO_8 \cdot 4H_2O$	100-400	
Goethite	α-FeOOH	8-200	Ni(II), Zn(II), Cd(II), Al(III), Cr(III), Ga(III), V(III), Mn(III), Co(III), Sc(III), Pb(IV), Ge(IV)
Hematite	α- Fe_2O_3	2-90	Al(III), Cr(III), Mn(III), Rh(III), Ga(III), In(III), Cu(II), Ge(IV), Sn(IV), Si(IV), Ti(IV)
Maghemite	γ- Fe_2O_3	8-130	Al(III)
Magnetite	Fe_3O_4 (FeO. Fe_2O_3)	4-100	Al(III), Mn(II), Ni(II), Cu(II), Co(II), Zn(II), Ca(II), Ge(IV)

Magnetite is often non-stoichiometric and both divalent and trivalent cations, such as Al, Cd, Co, Cr, Cu, Ni, Mn, and Zn, are easily integrated in its structure (Schwertmann and Cornell, 1991). The properties of some iron oxides are presented in Table 1.4.

1.4.6 Iron Oxide Coated Sand (IOCS)

Iron oxide coated sand is a by-product obtained in the treatment of groundwater containing iron. Iron oxide coating of the filter media is a natural process in which the coating develops in-situ within weeks to months on new or virgin filter media during filtration process. IOCS from different iron removal plants, being developed under different conditions, may have different physico-chemical characteristics and hence different adsorption capacities. Iron oxide coating in an iron removal treatment plant is affected by the raw water quality (e.g. pH, Mn, Fe, Ca, and TOC) that is being treated. It also depends upon the process condition applied, e.g. filtration rate, depth of the media, backwash conditions, etc. In the study conducted by Sharma (2001) IOCS from twelve different groundwater treatments plants were analyzed. It was found that IOCS has much high porosity (110 times) and very large specific area (5-200 times) compared to new virgin sand. Iron content of the coating ranged between 27% and 45 % and the coatings was not uniform (Sharma 2001).

1.4.7 Granular Ferric Hydroxide (GFH / GEH)

GFH is produced from a ferric chloride solution by neutralization and precipitation with sodium hydroxide. The ferric hydroxide precipitate is centrifuged and granulated by a high-pressure process. As no drying procedure is included in its preparation, all the pores are completely filled with water, leading to a high density of available adsorption sites and thus to a high adsorption capacity. The GFH consists of ferric oxihydroxide: approximately 52 to 57% by mass, while 43 to 48% is s moisture. GFH has grain porosity of 72 to 77% (Driehaus et al., 1998). The GFH used in this study is manufactured by GEH Wasserchemie GmbH made of akagenéite (β-FeOOH).

1.5 Adsorption theory

1.5.1 Introduction

Adsorption from solution onto a solid surface occurs as the result of one of the two characteristic properties for a given solvent-solute-solid system or a combination thereof. The first property is the driving force for adsorption (which is a consequence of the lypophobic character of the solute relative to the particular solvent), and the second one is a high affinity of the solute for the solid. For the majority of the water and wastewater treatment systems, adsorption results from a combined action of these two forces (Schippers *et al.*, 2007)

Adsorption process involves a quantitative equilibrium distribution between phases. Solutions containing known quantities of the substance of interest (adsorbate) are equilibrated with the adsorbent. At equilibrium, the amount of substance that has disappeared from the solution phase is assumed to be adsorbed by the solid. This solid solution distribution is then characterized by a distribution coefficient (Equation 1.1):

$$K_d = q/C_{eq} \tag{1.1}$$

Where q represents the equilibrium mass of adsorbed substance per unit mass of adsorbent, Ceq is the equilibrium mass of the substance in solution per unit volume of solution, and K_d represents a linear distribution coefficient and has units of volume per mass. Adsorption is described by chemical reactions involving specific chemical forms of the adsorbent and specific surface functional groups. Each reaction is further characterized by an equilibrium constant that can be applied to any environment, irrespective of the characteristics of the environment in which the adsorption reaction occurs (Faust and Aly, 1998).

1.5.2. Type of adsorption

Three different types of adsorption exist. These are ion exchange, physical adsorption and chemical adsorption. Exchange adsorption or ion exchange is a process in which ions of one substances concentrate at a surface as a result of electrostatic attraction to charged sites at the surface. For two potential ionic adsorbates, in the absence of other specific sorption effects, the charge of the ion is the determining factor for exchange adsorption (an ion with a high valency will be adsorbed faster). For ions of equal charge, molecular size determines order of preference for adsorption, the smaller being able to come closer to the adsorption site and thus being favoured. (Yang, 1999, Schippers *et al.*, 2007)

Physical adsorption is due to weak forces of attraction between molecules (Van Der Waals forces). The adsorbed material is not fixed to a specific site but is rather free to undergo translational movement with the interface. Physical adsorption is generally reversible (Schippers *et al.*, 2007).

Chemical adsorption, also termed chemisorption, takes place as a result of a chemical bond being formed between the molecule of the solute and the adsorbent. The adsorbed molecules are localized at specific sites and therefore are not free to migrate on the surface. Chemical adsorption is generally irreversible and exothermic (Schippers *et al.*, 2007). As an example, the adsorption of a metal ion on an oxide surface involves the formation of bonds of the metal ion with the surface oxygen atoms and the release of protons from the surface (Equation 1.2):

$$\equiv S - OH + M^{2+} \rightarrow \equiv S - OM^{+} + H^{+} \qquad (1.2)$$

where M^{2+} represents a divalent cation and S-OH represents an oxide surface. In the same way, anions adsorption by hydrous oxides occurs via ligand exchange reactions in which hydroxyl group are replaced by the sorbing ions (Dzombak and Morel, 1990):

$$\equiv S - OH + A^{2-} + H^{+} \rightarrow \equiv S - A^{-} + H_{2}O \text{ or} \qquad (1.3)$$

$$\equiv S - OH + A^{2-} + 2H^{+} \rightarrow \equiv SHA + H_{2}O \qquad (1.4)$$

where A^{2-} is the hypothetical divalent anion

1.5.3. Adsorption isotherms

Adsorption from aqueous solutions involves concentration of the solute on the solid surface. As the process proceeds, the sorbed solute tends to desorb into the solution. Equal amount of solute eventually are being adsorbed and desorbed simultaneously. Consequently, the rates of adsorption and desorption will attain an equilibrium state, called adsorption equilibrium. At equilibrium, no change can be observed in the concentration of the solute on the solid surface or in the bulk solution. The position of equilibrium is characteristic of the solute, adsorbent, solvent temperature and pH. The representation of the amount of solute adsorbed per unit of adsorbent (q) as a function of the equilibrium concentration of the solute in the bulk solution (C_e) at a constant pH and temperature is called an isotherm. Inspection of adsorption isotherms can provide the following valuable information (Schippers et al., 2007):

- The absorbability or relative affinity of a component for the adsorbent;
- The degree of removal achievable as indicated by the equilibrium adsorbate concentration;
- Sensitivity of the adsorbate concentration change as indicated by the relative steepness of the isotherm line;
- By obtaining adsorption isotherms at different pH, one can determine whether the added cost of pH adjustment will be offset by savings in the quantity of adsorbent used.

Several models can be used for the description of adsorption data with the Freundlich and Langmuir isotherms being most commonly used.

1.5.3.1 The Freundlich adsorption isotherm

The Freundlich adsorption isotherm is the most widely used mathematical description of adsorption in aqueous systems. The Freundlich equation (Equation1.5)is expressed as:

$$q = KC_e^{1/n} \tag{1.5}$$

where q = amount of solute adsorbed per unit weight of adsorbent = x/m (g/g)

C_e = equilibrium concentration of the solute (g/m^3)

K, $1/n$ = isotherm constants.

K is the measure of adsorption capacity and $1/n$ is the measure of adsorption intensity. Upon linearization, the equation takes the form (Equation 1.6):

$$log\ q = log\ K + 1/n\ log\ C_e \tag{1.6}$$

If $1/n$ is close to 1, this indicates a high adsorptive capacity at high equilibrium concentrations, which rapidly diminishes at lower equilibrium concentrations covered by the isotherm. Relatively flats slope, i.e. $1/n \ll 1$, indicates that adsorption capacity is only slightly reduced at the lower equilibrium concentrations. As the Freundlich equation indicates, the adsorptive capacity q is a function of the equilibrium concentration of the solute. Therefore, higher capacities are obtained at higher equilibrium concentrations (Faust and Aly, 1998).

1.5.3.2 The Langmuir adsorption isotherm

The Langmuir adsorption isotherm is perhaps the best known of all isotherms describing adsorption and is often expressed by Equation 1.7 (Casey, 1997):

$$Q_e = X_m K C_e / (1 + K C_e) \tag{1.7}$$

where:

- Q_e is the adsorption density at the equilibrium solute concentration (mg/g)
- C_e is the equilibrium concentration of adsorbate in solution (mg/l)
- X_m is the maximum adsorption capacity corresponding to complete monolayer coverage (mg of solute adsorbed per g of adsorbent)
- K is the Langmuir constant related to energy of adsorption (l of adsorbent per mg of adsorbate)

The above equation can be rearranged and linearized according to the Equation 1.8:

$$C_e/Q_e = 1/X_m K + C_e/X_m \tag{1.8}$$

The linear form can be used for linearization of experimental data by plotting C_e/Q_e against C_e. The Langmuir constants X_m and K can be evaluated from the slope and intercept of the linear equation.

The basic assumption underlying the Langmuir model, which is also called the ideal localized monolayer model are (Faust and Aly, 1998):

- the molecules are adsorbed on definite sites on the surface of the adsorbent;
- each site can accommodate only one molecule (monolayer);
- the area of each site is a fixed quantity determined solely by the geometry of the surface;

- the adsorption energy is the same at all sites;

1.5.3.3 The BET adsorption isotherm

Often molecules form multilayers, some are adsorbed on already adsorbed molecules and the Langmuir isotherm is not valid. In 1938 Stephan Brunauer, Paul Emmett and Edward Teller developed an isotherm that takes into account that possibility (Brunauer et al., 1938). Brunauer- Emmett- Teller (BET) equation (Equation 1.9) is as follow:

$$\frac{q}{q_m} = \frac{BC_e}{(C_e - C_s)[1 + (B-1)(C_e / C_s)]} \qquad (1.9)$$

Where B = is a dimensionless constant

C_s = saturation concentration of the adsorbate (g/m^3)
q = amount of solute adsorbed per unit weight of adsorbent (g/g)
q_m = maximum adsorption capacity (g/g)

The BET isotherm can be linearized as it is shown in the Equation 1.10:

$$\frac{C_e}{(C_s - C_e)q} = \frac{1}{Bq_m} + \frac{B-1}{Bq_m}\frac{C_e}{C_s} \qquad (1.10)$$

1.5.4 Factors affecting adsorption

Different factors affect the adsorption process. These are namely the characteristics of adsorbent, the nature of adsorbate, ionic concentration, organic matter, pH and temperature. The influence of these factors is discussed below.

The surface area is one of the principle characteristics affecting the adsorption capacity of the filter media, because the adsorption capacity of an adsorbent is proportional to the specific surface area, i.e., the adsorption of a certain solute increases with an increase of surface area (Faust and Aly, 1983). The surface area, per unit volume of non-porous adsorbent, increases with a decrease in particle size (Sharma, 2001). As a result the adsorptive capacity per unit weight of adsorbent increases with a reduction in particle diameter.

The composition of an adsorbent (especially iron oxides) might be useful preliminary information, when assessing suitability of adsorbent (IOCS, GFH) for heavy metal removal. The hydrated surface of oxides can adsorb cations and anions present in water. The extent of adsorption depends on the type and density of the adsorption sites available and the nature of the adsorbing ion (Sharma *et al.*, 2002)

Adsorption is influenced by several physico-chemicals properties of an adsorbate (substance to be adsorbed). These are solubility, surface charge, molecular weight, size of adsorbate molecule and ionic radii. Solubility is the most significant property affecting the adsorption capacity. The higher solubility indicates a stronger solute-solvent interaction of affinity and the extent of adsorption is expected to be low because of the necessity of breaking the solute-solvent interaction before adsorption can occur.

Other things being equal, the adsorption increases with atomic number and decreases with decreasing ionic size for transition metals (Sharma *et al.*, 2002).

Raw water contains a mixture of many ions and compounds rather than a single one. These ions may enhance adsorption, may act relatively independently or may interfere with one another. Adsorption of one single ion is sensitive to the ionic strength of the solution with adsorption capacity increasing by decreasing ionic concentration (Faust and Aly, 1983; Weber, 1972). Mutual inhibition can be predicted to occur if adsorption is confined to a single or a few molecular layers, the adsorption affinities of the solute do not differ by several orders of magnitude and there is no specific interaction between solutes enhancing adsorption. The degree of mutual inhibition is related to sizes, concentrations, and adsorption affinities of the competing molecules (Weber, 1972).

Ligands, mostly anions, can affect adsorption of metal ions onto oxides surfaces in four ways (Sharma et al, 2002):

-   Metal-ligand complexes may form in solution and adsorb only weakly or not at all
-   The species may interact indirectly at the surface, thereby altering the surface electrical properties
-   The metal-ligand complex may adsorb strongly, thereby enhancing the adsorption of the metal-ions
-   The formed complex may have no effect on metal adsorption

Hence, depending on the nature of the anions present and complexes formed, metal adsorption onto iron oxide surface (IOCS or GFH) may be enhanced, decreased or unaltered.

The organic matter (OM) that is of natural origin is derived primarily from plants and / or microbial residues. In their original or chemically modified form, the residues of organic matter produced on land are available to be transferred from the soil into the hydrosphere. Transport usually occurs due to rainfall that runs off or percolates through the soil column carrying dissolved organic matter (DOM) and particulate organic matter (POM) to streams, lakes, and oceans or into groundwater. Humic substances (HS) which are a form of environmental organic matter of plant or microbial origin, besides playing a role as a proton acceptor and contributing to charge balance in aqueous systems, also react with metals in solution through the formation of ionic or covalent bonds (Snoeyink and Jekins, 1980). The formation of one or another of these associations between humic acid and metals depends on the initial state of the humic substance and the metal and their concentrations. Humic substances show ion exchange and complexing properties with most constituents of water thus influencing the water treatment processes (Sharma, 1997).

The size and the polarity of organic matter also affect adsorption. In general, the solubility of any organic compound in water decreases with increasing chain length or increasing size of the molecule (Faust and Aly, 1983).

Since most of organic compounds in water are usually dissolved, their presence considerably affects the adsorption of metals present. In general when more than one adsorbable component is present, and adsorption sites are limited, competitive adsorption occurs and the adsorption of some ions may be limited by the lack of active sites (Salomon and Forstner, 1995, Sharma, 1997).

Typically, there is a dramatic increase or decrease in adsorption of cations and anions as pH increases through a critical pH range. It has been observed that, for a given adsorbate/adsorbent ratio, there is a narrow pH range, over which the adsorption of cations and anions of hydrous oxides increases from 0 to 100 %, giving typical pH versus percentage adsorption curves known as adsorption pH edges. As the adsorbate/adsorbent ratio is increased, the fractional adsorption at a given pH is reduced and consequently the cation "pH edge" shifts to the right (Abdus-Salam, 2005).

As the pH increases, a dramatic adsorption of cations is observed while the adsorption of anions decreases. Studies on Pb and Cd adsorption onto goethite showed the expected trend of increasing metal retention with increasing medium pH. When the pH of the adsorbing medium is increased from pH 3–5, there was a corresponding increase in deprotonation of the goethite surface leading to a decrease in H^+ ion on the goethite surface. This creates more negative charges on the goethite surface, which favors adsorption of positively charge species as a result of less repulsion between the positively charge species and the positive sites on the goethite surface (Abdus-Salam, 2005).

1.6 Use of iron oxides based media in the removal of heavy metals

The iron oxides based adsorbents provide well binding properties for heavy metals that can either sorb to surfaces, or substitute for Fe in the bulk structure. This capability of iron oxides has been widely used in removing heavy metals from contaminated environment (Damirrel et al., 1999, Martin and Kempton, 2000).

The adsorption of lead (II) and zinc(II) onto goethite was studied as a function of pH, total dissolved metal concentration, surface area of goethite, and ionic strength. The adsorption edge of lead ranged from pH 4 to 7, similar to that of copper, but the adsorption edge of zinc was displaced by 1.5 pH units toward higher pH (Kooner, 1993). The influence of pH and adsorbent concentrations on the sorption of Pb and Cd by the natural goethite was also studied by Abdu Salam and Adekola (2005). Sorption efficiency was strongly governed by pH, with nearly 100% adsorption of Pb occurring at an initial pH of 5. Generally, Pb was sorbed more efficiently than Cd with increasing pH from 3 to 5. Efficient removal of Pb was achieved at a lower dose of goethite, as there was no appreciable increase in the amount of Pb adsorbed when the adsorbent dose was increased (Abdu-Salam and Adekola, 2005).

Due to its ferromagnetic property, natural magnetite can be used not only as an adsorbent for removing metals but also as a magnetically energizable element for attracting and retaining paramagnetic nanoparticles, thus removing them from solution. Synthetic magnetite has been found to be an effective sorbent for heavy metal ions including Cu(II), Cd(II) and Pb(II) (Shanaka De Silva, 2007).

A number of studies have shown that lead can be adsorbed on metal oxides, particularly goethite (Forbes et al., 1976), sand or other materials coated with manganese dioxide (Degs et al., 2001). Lead can also be adsorbed on iron or aluminium oxihydroxide deposited on sand (Lai, 2000), on granular iron oxihydroxide (Theis et al., 1992), or on activated alumina. Iron oxides gave better results than aluminium oxides.

Adsorption of iron was found to be 20-25 times higher on iron oxihydroxide-coated sand than on sand with no coating, and this is therefore a very good treatment alternative for removal of divalent iron from ground water. After adsorption of iron II, the latter is oxidized and in turn forms ferric-hydro-oxide (Sharma et al., 1997, Sharma et al., 2003).

The adsorption of iron is accompanied by the release of H^+ ions. But if the pH of the water is lowered, the quantity of iron adsorbed decreases. This is shown by the Equations 1.11-1.13

$$\equiv S - OH + Fe^{2+} \leftrightarrow \equiv SO\,Fe^+ + H^+ \tag{1.11}$$

$$\equiv SO\,Fe^+ + \frac{1}{4}\,O_2 + 3/2\,H_2O \leftrightarrow \equiv S-O\,Fe\,(OH)_2 + H^+ \tag{1.12}$$

$$\equiv S - OH + Fe^{2+} + \frac{1}{4}\,O_2 + 3/2\,H_2O \leftrightarrow \equiv S-O\,Fe\,(OH)_2 + 2H^+ \tag{1.13}$$

Depending on pH, iron oxihydroxides can adsorb arsenate and arsenite (Manning and Goldberg, 1997). At a pH close to neutrality, both forms of arsenic are retained (Huang, 1989). Recently studies have shown efficient removal of As(III) and As (V) with IOCS media (Sharma 1997, Petrusevski et al. 2001).

Granular ferric hydroxide (GFH) can also remove very effectively arsenic from ground water. Various tests conducted with the adsorbents have shown a high adsorption capacity of the adsorbent (40, 000 to 60, 000 bed volumes), until the permissible limit for arsenic (10 μg/L) was exceeded. It was found that the adsorption capacity of GFH is 5 to 10 times higher than that of activated Alumina. The typical residual mass of the spent GFH was in the range of only 5 to 25 g/m^3 treated water, whereas that of the spent Activated Alumina was 10 times higher. It was also found that under normal environment conditions, no leaching of arsenic takes place out of spent GFH (Pal, 2005).

Bailey et al., 1992 showed that iron oxihydroxide can retain 99 % of Cr(VI). Studies on adsorption competition have shown that the presence of As(V) reduces the adsorption of chromate (Khaodhiar et al., 2000). Recent studies on removal of Cr with IOCS gave promising results. Freundlich isotherm constants K for Cr(III) and Cr(VI) were found to be 54.73 mg/g and 0.299 mg/g, respectively (Tessema, 2004).

Several studies have been conducted on the removal of heavy metals by iron oxides based media. Tekeste (2003) conducted batch experiments on the removal of Cd, Cu, Cr, Ni and Pb by IOCS. He showed that IOCS has very high adsorption capacities for the heavy metals studied. Under the given conditions, over 90 % of the metals studied were removed from model groundwater after 5 days of contact time. Another study conducted by Devendra (2007) on the removal of Cu, Cd, Cr(III) and Cr(VI) from urban storm water run-off using IOCS and GFH also gave promising results. The results have shown that IOCS is capable of removing all heavy metals studied while GFH was capable of removing Cu, and Cr but not Cd.

1.7 Rapid Small Scale Column Test

Recent research has developed a method that can help to rapidly determine adsorption media life in full scale treatments plants. This bench scale method, called the rapid small scale column test (RSSCT), is performed in a laboratory. Through a process called dimensional scaling, RSSCT can simulate full-scale performance in approximately $1/10^{th}$ the duration required for pilot test, with only a fraction of the water volume (Brandhuber, 2004). There are three primary advantages in using RSSCT for design: i) an RSSCT may be conducted in a fraction of the time that is required to conduct pilot studies; ii) unlike predictive mathematical models, extensive isotherm or kinetic studies are not required to obtain a full scale performance prediction from RSSCT; and iii) a small volume of water is required to conduct the test. Consequently, replacing a pilot study with a RSSCT could significantly reduce the time and costs of pilot investigations that are required for a full-scale design (Crittenden et al., 1986). RSSCT approaches are believed to be a powerful method not only to predict full-scale but also as a research tool (Sperlich et al., 2005).

Mathematical models are used to scale-down a full-scale adsorptive filter column to an RSSCT and to maintain perfect similarity between performances of the full- and small scale adsorptive filter column.

The relationship between small- and full-scale column design run time is defined by ratio of media particle diameters by the Equation 1.14:

$$\frac{d_{P_{SC}}}{d_{P_{LC}}} = \frac{EBCT_{SC}}{EBCT_{LC}} = \frac{t_{SC}}{t_{LC}} \tag{1.14}$$

Where d_p = media diameter, EBCT = Empty Bed Contact Time, t = run duration, SC = small column, LC = large column

This equation shows that the ratio of small- and full-scale column run time (time to breakthrough) is proportional to the relative diameter of the treatment media. By knowing the relative diameter of the treatment media in the small- and full-size column, and by determining the run time in the small column by RSSCT, the full-scale performance can be calculated (Brandhuber, 2004).

Perfect similarity between RSSCT and full-scale adsorptive filter columns may be achieved if the surface diffusivities, isotherm capacities, bulk densities, operating temperature and influent concentrations are the same (Crittenden et al., 1986). Accordingly, scaling procedures were developed that assume the surface diffusion is the controlling mechanism, because the effects of other mechanisms are usually small compared to surface diffusion (Crittenden et al., 1986). Scaling can be based on upon proportional diffusivity (PD) or constant diffusivity (CD). The PD approach assumes that the effective surface diffusivity is linearly proportional to the particle radius, and surface diffusion is the controlling mechanism while the CD approach is used when the effective surface diffusivity is independent of particle size, and hence identical for full-scale and RSSCT columns (Sperlich et al., 2005).

For a number of years, the RSSCT procedure has been widely accepted as a mean to quickly and inexpensively provide estimates of full-scale GAC treatment performance. Due to the similarity in adsorptive arsenic media and the GAC treatment process, much work has been performed to validate the RSSCT method for arsenic treatment. RSSCT tests for arsenate removal by GFH from model and groundwater has been conducted; adsorption kinetics studies and a comparison of laboratory RSSCT performance versus pilot-scale performance suggests that proportional diffusivity (PD) RSSCT scaling approaches are more valid than constant diffusivity (CD) approaches for arsenate adsorption onto GFH (Westerhoff et al., 2005), but it was concluded that more pilot-scale data are necessary to confirm the results. RSSCT tests have also been conducted in the removal of heavy metals from urban storm water run-off by IOCS and GFH media. The results obtained were in agreement with the results from batch experiments (Devendra, 2007).

Because of its speed and simplicity, there are numerous potential applications of the RSSCT. In addition to using the technique directly comparing the arsenic removal performance of disposable media, RSSCT can be used to (Brandhuber, 2004):

- Optimize treatment pH
- Optimize hydraulic loading rate and EBCT
- Evaluate the effect of intermittent treatment in operation
- Study impact of varying water quality and blend scenarios
- Predict changes to treated water over time
- Perform media quality control

1.8 Scope of the thesis

This thesis presents the work done on the removal of different heavy metals by iron oxides based media (IOCS and GFH) at laboratory scale. The research aimed at investigating the adsorption process involved in heavy metals removal from groundwater. Major mechanisms of metal removal are explained on the basis of experimental results and available information in the literature. The work was done by conducting laboratory batch scale experiments at the National University of Rwanda (NUR) and UNESCO-IHE in Delft, The Netherlands. Besides batch experiments, the Rapid Small Scale Column Test (RSSCT) was used in order to predict adsorption capacity and performance of full-scale adsorptive filters based on IOCS and GFH. As groundwater in Rwanda is still unexplored field, Chapter 2 mainly focused on the groundwater quality in Rwanda. Screening of groundwater quality was conducted in the Eastern province of Rwanda, where groundwater is the main source of drinking water. Chapters 3, 4, and 5 show the effects of water matrix on the adsorptive removal of different heavy metals. Chapter 3 shows the effect of Calcium on As(III) and As(V) removal, Chapter 4 focuses on the competitive effect of phosphate on Cr(VI) removal, while Chapter 5 looks at the competitive effects of Ca on Cd and Cu removal by IOCS and GFH. Chapter 6 presents the results of the effects of organics (fulvic acid -FA) on the adsorptive removal of As(V) and Cr(VI) by IOCS and GFH. Chapter 7 focuses on describing the sorption reactions using the surface complexation modelling. Finally, the thesis ends with a summary of the study and outlook.

1.9 References

Abdu-salam N., Adekola, F. A., 2005 The influence of pH and adsorbent concentration on adsorption of Lead, and Cadmium on natural goethite, *African Journal of Science and Technology (AJST)Science and Engineering Series Vol. 6, No. 2, pp. 55 – 66*

Adepoju-Bello, A.A. and Alabi O.M., 2005 Heavy metals: A review. The Nig. J. Pharm., 37: 41-45.

Adepoju-Bello, A.A., Ojomolade O.OAyoola.G.A. and Coker H.A.B., 2009, Quantitative analysis of some toxic metals in domestic water obtained from Lagos metropolis. The Nig. J. Pharm. 42(1): 57-60.

Adewole AT (2009). Waste management towards sustainable development in Nigeria: A case study of Lagos state. Int. NGO J. 4(4): 173-179.

Adeyemi, O., O.B. Oloyede and A.T. Oladiji, 2007, Physicochemical and microbial characteristics of Leachate contaminated ground water. Asian J. Biochem., 2(5): 343-348.

AFSSA, (2005) Evaluation of the use of metal oxide-coated sands for the treatment of water for human consumption and natural mineral water.

Aziz H.A., Yusoff M.S., Adlan M.N., Adnan N.H. and Alias S., Waste Manag., 24 (2004) 353– 358.

Bakare-Odunola, M.T., 2005. Determination of some metallic impurities present in soft drinks marketed in Nigeria. The Nig. J. Pharm., 4(1): 51-54.

Bailey R.P., Bennett T., Benjamin M.M. 1992, Sorption onto and recovery of Cr (VI) using iron oxide coated sand. *Wat. Sci. Technol.* 26(5-6), pp. 1239-1244.

Baldwin R. D. and Marchall J.W. 1999, Heavy metals poisoning and its laboratory investigation, *Ann. Clin. Biochem,* 36: 267-300

Benjamin M.M. 1983, Adsorption and surface precipitation of metals on amorphous iron oxihydroxides. *Environ Sci. Technol.* 17 (11), pp. 686-691.

Brandhuber P., (2005), Simulating arsenic treatment Media performance Through Rapid Small Scale Column Testing. *Waterscapes*, volume 16, number 4.

Brunauer S., Emmett P.H., Teller E., 1938, *J. Am. Chem. Soc.*, 60, 309.

CASEY, T.J., 1997, Unit Treatment Processes in Water and Wastewater Engineering , John Wiley and Sons Ltd, England, pp113-114

Chin H.L. 2002. Cadmium adsorption on goethite-coated sand in the presence of humic acid, water Research, 36 (20), 4943-4950

Crittenden J.C., Berrigan J.K. and Hand D.M. 1986, Design of rapid small scale column adsorption tests for a constant diffusivity, *Journal of water pollution control federation* 58 (4), 312-319

Degs Y.A., Khraisheh M.A.M., Tutunji M.F. 2001, Sorption of lead ions on diatomite and manganese oxides modified diatomite. *Wat. Res.* 35, n° 15, pp. 3724-3728.

Demirel, B., Yenigün, O., and Bekbölet, M. (1999): Removal of Cu, Ni and Zn from wastewaters by the ferrite process. *Environ. Technol.*, 20:963-970

Devendra Y. (2007), Adsorptive removal of heavy metals from urban storm water run-off, Msc. Thesis, Unesco-IHE, Delft

Driehaus, W., Jekel, M., Hildebrandt, U., 1998. Granular ferric hydroxides e A new adsorbent for the removal of arsenic from natural water. J. Water SRT-Aqua 47 (1), 30-35.

Dzombak D.A. and Morel F.M.M., 1990, Surface Complexation Modeling: Hydrous Ferric Oxide. John Wylie and Sons, New York.

Faust, S.D., Aly O.M., (1998), Chemistry of water Treatment, second edition, Lewis publisher, Boca Raton

Forbes E.A., Posner A.M., Quirk J.P. 1976, The specific adsorption of divalent Cd, Co, Cu, Pb and Zn on goethite. *J. Soil Sci.* 27, pp. 154-158.

Gerth, J. 1990: Unit-cell dimensions of pure and trace metal-associated goethites. *Geochim. Cosmochim. Acta.*, 54:363-371.

Hammer, M.J., and Hammer, Jr., 2004. Water Quality. In: Water and Waste Water Technology. 5th Edn. New Jersey: Prentice-Hall, pp: 139-159.

Hu H. , 1998, Chapter 397: Heavy metal poisoning. In: Fauci AS, Braunwald E, Isselbacher KJ,Wilson JD, Martin JB, Kasper DL, Hauser SL, Longo DL (eds). Harrison's principles of Internal medicine. 14th ed. New York: McGraw-Hill; pp2564-2569.

Huang P. M. 1989, In minerals and soil environment SSSA Book ser.no1, pp 975-1050

Igwilo, I.O., O.J. Afonne, U.J. Maduabuchi and O.E. Orisakwe, 2006. Toxicological study of the Anam River in Otuocha, Anambra State, Nigeria. Arch. Environ. Occup. Health, 61(5): 205-208.

Jambor, J.L. and Dutrizac, J.E. 1998: Occurrence and constitution of natural and synthetic ferrihydrite, a widespread iron oxyhydroxide. *Chem. Rev.*, 98:2549-2585

James R.O. and Ramamoorthy S. 1984, Heavy metals in natural waters: Applied monitoring and impact assessment, New York, USA

Karaschunke K., and Jekel M., 2002, Arsenic removal by iron hydroxides by enhanced corrosion of iron, *Water Supply*, Vol. 2, No 2, pp 237–245

Khaodhiar S., Azizian M.F., Osathapan K., Nelson P.O., 2000, Copper chromium and arsenic adsorption and equilibrium modelling on iron-oxide coated sand, background electrolyte system, *Water Air Soil Pollution* 119, pp. 105-120.

Kooner Z.S. 1993, Comparative study of adsorption behavior of copper, lead, and zinc onto goethite in aqueous systems, *Environmental geology*, Volume 21, number 4, 242-250,

Lai C.H., Chen C.Y., Shih P.H., Hsia T.H., 2000, Competitive adsorption of copper and lead ions on iron-coated sand from water .*Wat. Sci. Technol.* 42 (3-4), pp.149-154

Lee G.F., Lee J.A. 2005. Municipal solid waste landfills – water quality issues, Water Encyclopaedia: Water Quality and Resource Development, John Wiley. NJ pp. 163-169

Lo S.L., Jeng H.T., Lai C.H., 1997, Characteristics and adsorption properties of iron-coated sand *Wat. Sci. Tech.*, 35 (7), pp. 63-70

Manning B.A., Goldberg S., 1997, Adsorption and stability of arsenic(III) at the clay mineral-water interface *Environ. Sci. Tech.* 31, pp. 2005-2011

Marcovecchio, J.E., S.E. Botte and R.H. Freije, 2007. Heavy Metals, Major Metals, Trace Elements. In:Handbook of Water Analysis. L.M. Nollet, (Ed.). 2nd Edn. London: CRC Press, pp: 275-311.

Martin, T.A. and Kempton, J.H. (2000): In situ stabilization of metal-contaminated groundwater by hydrous ferric oxide: An experimental and modeling investigation. *Environ. Sci.Techn.*, 34:3229-3234.

McMurry, J. and R.C. Fay, 2004. Hydrogen, Oxygen and Water. In: McMurry Fay Chemistry.

K.P. Hamann, (Ed.). 4th Edn. New Jersey: Pearson Education, pp: 575-599.

Mendie, U., 2005. The Nature of Water. In: The Theory and Practice of Clean Water Production for Domestic and Industrial Use. Lagos: Lacto-Medals Publishers, pp: 1-21.

Momodu M.A. and Anyakora C.A. 2010, Heavy Metal Contamination of Ground Water: The

Surulere Case Study, Research Journal Environmental and Earth Sciences 2(1): 39-43

Mull EJ 2005. Approaches toward Sustainable Urban Solid Waste Management: Sahakaranagar, layout, Unpublished M.Sc. Int.Environ. Sci., Lund University, Lund, Sweden p. 37

Pal B.N., 2001, Granular Ferric Hydroxide for Elimination of Arsenic from Drinking Water, *BUET-UNU International workshop on Technology for arsenic removal from drinking water,* 5-7 May, p 59-68

Pertusevski B., Boer, J., Shahidullah, S.M., Sharma, S.K., Schippers, J.C., 2000, Adsorbents based point of use system for arsenic removal in rural areas, proceeding of conference on innovation in conventional and advanced water treatment processes, Amsterdam

Salomons W. and Forstner U. 1995, Heavy metals, problems and solutions, Springer-Verlag, Berlin, New York

Sampat, P. 2000, Groundwater shock, World Watch, January/February, 10-22

Schippers J.C., Petrusevski B, Sharma S.K. Amy G.L., 2007, Module groundwater resources And Treatment, Unesco-IHE, Delft

Schwertmann U, Cornell RM. 1991.Iron oxides in the lab- oratory.Weinheim: VCH Verl. p 117

Shanaka De Silva K. (2007), studies of magnetic filtration techniques to purify potable water and waste water , Msc thesis, Massey University, New Zeland

Sharma R.S. and Al-Busaidi T.S., 2001 Groundwater pollution due to a tailings dam/ Eng. Geol., 60 235–244.

Sharma S.K. 2001 Adsorptive iron removal from Groundwater, PhD thesis, Unesco-IHE, Delft

Sharma S.K., Petrusevski B., Schippers J.C., 2002, Characterization of coated sand from iron removal plants. *J. Water Sci. Technol. Water Supply* 2.2, pp. 247-257

Sharma S.K., Pretusevski B., Heijman B., Schippers J.C., 2003, Prediction of iron(II) breakthrough in adsorptive filters under anoxic conditions. *J. Water Supply Res. Technol.* Aqua 52.8, pp. 529-544

Shiklomanov, A. & Rodda J. C. 2003 Summary of the Monograph 'World Water Resources at the beginning of the 21st Century', prepared in the framework of IHP UNESCO, 1999.Vie

Singh, B. and Gilkes, R.J. 1992: Properties and distribution of iron oxides and their association with minor elements in the soils of south-western Australia. *J. Soil Sci.,* 43:77-98.

Sloof W., Cleven R.F.M., Janus J.A., Van der Poel P., 1990, Integrated criteria document, Chromium, National Institute of Public Health and Environmental Protection (RIVM) Bilthoven, Netherlands, report number 710401002

Snoeyink, V.L. and Jenkins, D. 1980 Water chemistry, John Wiley & SOns, USA

Sørensen M.A., 2001 Iron Oxides as a Stabilizing Agent for Heavy Metals, Ph.D. Thesis, *Environment & Resources DTU* Technical University of Denmark

Sorg, T.J., 1986, Removing dissolved inorganic contaminants from water. Third of a six-part series on water treatment processes. *Environ. Sci. and Tech.* 20 (11)

Sperlich A., Werner A., Genz A., Amy G., Worch E., Jekel M., 2005, Breakthrough behavior of granular ferric hydroxide (GFH) fixed bed adsorption filters: modeling and xperimental approaches, *Water research* 39 (6), 1190-1198

Stumm W., 1992, Chemistry of the solid-water interface, John Wiley & Sons, Inc

Stumm W., Morgan J.J., 1996, Aquatic chemistry, New York, NY, Wiley Interscience

Tekeste E.A. 2003, Removal of heavy metals from groundwater using Iron Oxide Coated

Sand, (Msc Thesis), Unesco-IHE, Delft

Tessema S.T., 2004, Removal of Chromium from groundwater using IOCS, M.Sc. Thesis, Unesco-IHE, Delft

Theis T.L., Iyer R., Ellis S.K., 1992, Evaluating a new granular iron oxide for removing lead from drinking water. *J. Am. Wat. Works Association* 84, pp. 101-105.

Vaishya R.C., Gupta S.K., 2003, Coated sand filtrations an emerging technology for water treatment. *J. Water Supply Research Technol. Aqua* 52-4, pp. 299-305

UNEP, 2003 Groundwater and its Susceptibility to Degradation, UNEP/DEWA, Nairobi

Vanloon, G.W. and S.J. Duffy, 2005. The Hydrosphere, In: Environmental Chemistry: A Global Perspective. 2nd Edn. New York: Oxford University Press, pp: 197-211.

Vodela, J.K., J.A. Renden, S.D. Lenz, W.H. Mchel Henney and B.W. Kemppainen, 1997 Drinking water contaminants. Poult. Sci., 76: 1474-1492.

Vrba, J., 1985 Impact of domestic and industrial wastes and agricultural activities on groundwater quality. In: Hydrogeology in the service of man, Vol. XVIII, Part I, pp.91 – 117, IAH Memoires of the 18th Congress, Cambridge

Vrba J.and Gun J., (2004) The world's groundwater resources Contribution to Chapter 4 of WWDR-2, International Groundwater Resources Assessments Centre, Report Nr IP 2004-1

Weber, W. J. Jr. 1972, Physicochemical Processes. Wiley-Interscience, New York,

Westerhoff D, Highfield, M., Badruzzaman and Yoon Y, 2005, Rapid small-scale column tests for arsenate removal in iron oxide packed bed columns, *ASCE J, Environ. Eng.* 131 (2), pp 262-271

WHO, 1996, *Guidelines for drinking water quality*, Second edition, volume 2, Health criteria and other supporting information

WHO, 2004, *Guidelines for drinking water quality*, volume 1, Recommendations, 3rd ed., Geneva, Switzerland, World Health Organization

WHO, 2007, Water for Pharmaceutical Use In: Quality Assurance of Pharmaceuticals: A Compendium of Guidelines and Related Materials, 2nd Updated Edn.World Health Organisation, Geneva, 2: 170-187.

Yang R.T., 1999, Gas separation by Adsorption process, Serie of chemical enginnering, Vol.1, Publishers: Imperial College Press

Chapter 2: Assessment of groundwater quality in Eastern Rwanda; case study of Nyagatare District

Abstract

Surface water is the main source of drinking water and groundwater represents only 10 % of the total drinking water produced in Rwanda. However supply of drinking water is still inadequate and clearly groundwater is needed to supplement surface water sources. Groundwater in Rwanda also remains unexplored field and very limited information is available on the quality of this source of drinking water. The objective of Chapter 2 is to discuss the major water quality aspects for wells in the Nyagatare district that were covered by this study. The groundwater samples were collected from 20 boreholes in the Nyagatare District and 27 parameters were analysed. Five out of 27 parameters had values below detection limits and 22 parameters were considered for further analyses. Chemical analyses were carried out for the major ion concentrations of the water samples collected from different locations using the standard procedures recommended by APHA (1994). In this study, Piper trilinear diagram was used to classify the groundwater of the Nyagatare District using the USGS groundwater chart. Factor analysis, using SPSS version 18.0, was primarily used for data reduction. The factor analysis was carried out. The Piper Trilinear diagram showed that most of sampled sites are mainly sodium and potassium type and for few of wells, there is no dominant groundwater type. In terms of anions, few sites have chloride groundwater type, one has bicarbonate groundwater type and others have no dominant anions. The extracted components explain nearly 94% of the variability in the original 22 variables, and one can considerably reduce the complexity of the data set by using these components, with only a 6% loss of information. Six components for Eigen values greater than 1 were extracted. The first component is most highly correlated with fluoride, pH and sulfate (with corresponding correlation factors of 0.931, 0.922 and 0.914, respectively). The second component is most highly correlated with calcium and total hardness (corresponding correlation factors of 0.914 and 0.910, respectively) and the third component is most highly correlated with total alkalinity (0.616). The fourth, fifth, and sixth components are mostly correlated with potassium, iron and magnesium, respectively. Regarding the hardness of Nyagatare groundwater, 7 samples fall under soft class, 3 samples fall under moderately hard class, 7 samples fall under hard and 3 samples fall under very hard class. The calculation of percentage of Na^+, RSC and SAR showed that Nyagatare groundwater is suitable for irrigation. Nyagatare District having abundant granite and granite rocks being igneous rocks, this can explain the source of fluoride found in groundwater. The source of EC, TDS, ammonia and nitrite in Nyagatare groundwater can be related to human activities by application of fertilizers and manure.

2.1 Introduction

Water quality requirements can be usefully determined only in terms of suitability for purposes, or in relation to the control of defined impacts on water quality. For example, water that is to be used for potable purposes should not contain any chemicals or microorganisms that could be hazardous to health. Similarly, water for agricultural irrigation should have low sodium and boron content, while that used for steam generation and related industrial uses should be low in certain other inorganic chemicals. Water quality data are also required for pollution control, and the assessment of long-term trends and environmental impacts. The composition of surface and underground waters is dependent on natural factors (geological, topographical, meteorological, hydrological and biological) in the drainage basin and varies with seasonal differences in runoff volumes, weather conditions and water levels. Large natural variations in water quality may, therefore, be observed even where only a single watercourse is involved.

Water quality analysis is one of the most important aspects in groundwater studies. A hydro chemical study reveals the quality of water that is suitable for drinking, agriculture and industrial purposes. Further, it is possible to understand the change in quality due to mineral-water interaction or any type of anthropogenic influence (Sadashivaiah et al. 2008). Groundwater often contains seven major chemical elements (ions): Ca^{+2}, Mg^{+2}, Cl^-, HCO_3^- Na^+, K^+, and SO_4^{-2}. The chemical parameters of groundwater play a significant role in classifying and assessing water quality. Chemical classification also reveals the concentration of various predominant cations, anions and their interrelationships. A number of techniques and methods have been developed to interpret such chemical data. Zaporozee (1972) has summarized the various modes of data representation and has discussed their possible uses.

Groundwater in Rwanda remains unexplored field and very limited information is available on the quality of this source of drinking water. Thus, the objective of the present work is to inventory the major ion chemistry of groundwater of the Nyagatare District, in the Eastern province of Rwanda. The Nyagatare district was chosen because drilling projects have been conducted in the area. Piper diagrams and principle component (statistical) analysis (PCA) are used as methods of analysis.

2.2 Study area

Rwanda is a Sub-Saharan developing country where water scarcity is an issue. Although Rwanda has enough surface and sub-surface water, more than half of the population has no access to improved water supply systems. Nearly 85% of the used water sources are contaminated (Osodo and Rwamugema, 2001). Rwanda territories contribute to the discharge of two large basins in Africa. The first is the Nile basin that occupies 76% of the surface area and drains 10% of the total stream flow from the country through the principal tributaries of the Lake Victoria. The second basin is the Congolese basin occupying 24% of the country and draining 10% of the national waters from Lake Kivu to Lake Tanganyika (FAO, 2005).

In Rwanda, groundwater is an unexplored water resource. However, some drilling projects have been conducted all over the country especially in the Eastern Province and a limited number of approximately 22 000 sources have been inventoried in the country until now. Poor water management and lack of reliable information in the water sector has made it impossible to keep track of the exploited groundwater resources in Rwanda (FAO, 2005). But the very important erosion of the basin slopes does not support a normal recharge of the groundwater. With the available information from some projects, it is estimated that the total recharge throughout the country is 66 m^3/s. Of that discharge, there are 22 recognized springs which have a discharge of 9.0 m^3/s, but the population is only consuming only 0.9 m^3/s and the rest of the water is not used (AQUASTAT, 2005, Baligira, 2007 and Kabalisa, 2006). According to MINITERE-ISAR (2007), the geological formations of Rwanda are mainly composed of plio-quaternaly formations. This geological structure is mainly composed of alluvia that are situated in the hollows of the rivers, volcanic formations located in volcanic areas of the North and in the Southwest, the granite and associated rocks covering the healthy granite are almost found everywhere in the country as well as schist associated with quartzite.

Nyagatare, the largest district in Rwanda, lies in an area of grassy plains, and low hills, with excellent views in all directions, including the mountains of southern Uganda and, on a very clear day, the Virunga volcano range. The District of Nyagatare experiences small quantities of rain and hot temperatures. It is characterized by two main seasons: one long dry season that varies between 3 and 5 months with an annual average temperature varying between 25.3°C and 27.7°C. The monthly distribution of the rains varies from one year to another. Annual rainfall is both very weak (827 mm) and very unpredictable, making it difficult to satisfy the needs of agriculture and livestock. The hydrographic network is very limited in the District of Nyagatare. A part from the River Muvumba cuts across the District, and the Akagera and Umuyanja Rivers pass the District constituting its limits with Tanzania and Uganda, respectively; there is no other consistent river that can be exploited by the population in Nyagatare. The few rivers that are found there such as Nyiragahaya, Kayihenda, Kiruruma, Nayagasharara and Kaborogota are erratic and intermittent. The weak river network constitutes a serious handicap to responding to the needs of water for people and animals. Figure 2.1 shows the location of Nyagatare District on the Rwanda map as well as the location of boreholes.

2.3 Methodology

The groundwater samples were collected from 20 boreholes among 91 boreholes found in the Nyagatare District. The sampled boreholes were chosen randomly in different sectors where boreholes are found: 1 borehole was sampled in Musheli, 2 in Matimba, 3 in Rwempasha, 3 in Rwimiyaga, 4 in Karangazi, 2 in Nyagatare, 2 in Katabagemu, 1 in Rukomo and 2 in Tabangwe Sector. In this study, 27 parameters were analysed for all taken samples and some parameters were analysed onsite. These are temperature (T), pH, turbidity, conductivity, dissolved oxygen (DO) and Fe^{2+}. Other parameters were analysed in the laboratory from samples collected in different locations using the standard procedures recommended by APHA (1994)., These are total alkalinity (TA), total hardness (TH), total dissolved solids (TDS), Ca, K, Na, Mg, NH$_3$, NO$_3^-$, NO$_2^-$, Cl$^-$, F$^-$, PO$_4^{3-}$, SO$_4^{2-}$, Zn^{2+}, Mn^{2+}, Pb^{2+}, Cd^{2+}, Cu^{2+}, As and Cr. In this study, the Piper Trilinear diagram was used to classify the groundwater of the Nyagatare district using the USGS groundwater chart. The factor analysis was also used to reduce the data and was carried out using SPSS software, version 18.0. Five parameters were excluded in the analysis because their concentrations were below the detection limits including Pb^{2+}, Cd^{2+}, Cu^{2+}, As and Cr. Factors were extracted using Varimax with Kaiser normalization. The Cattel's scree test (Stevens, 1996, Kim and Mueller, 1978) was used to determine the number of factors to be extracted.

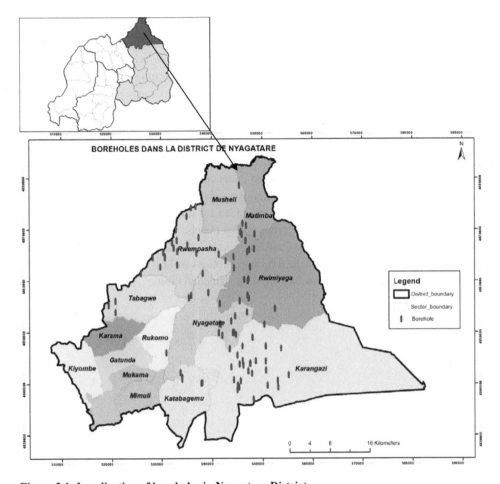

Figure 2.1: Localization of boreholes in Nyagatare District

2.4 Results

2.4.1 Physico-chemical characteristics of Nyagatare groundwater

The results obtained for different sites are presented in Table 2.1. The temperature varied depending on the time of sampling. pH values respect the norms of drinking water (6.5-8.5) except for one site in Nyagatare (Nyagatare as cell not as District) and all sampled sites in Katabagemu (2 sites), Rukomo (1 site) and Tabagwe (2 sites) where water is acidic. Two sites in Karangazi cell also showed high pH values (8.2 and 9.7) indicating that groundwater in those sites is probably polluted. Turbidity and conductivity for all sampled sites are within the range of acceptable values for drinking water. The dissolved oxygen is very low indicating that Nyagatare groundwater is anoxic. There is no WHO based guideline for the Total Dissolved Solids (TDS) but drinking-water becomes significantly and increasingly unpalatable at TDS levels greater than about 1000 mg/l (WHO, 2011). For all sampled sites, TDS values are less than 1000 mg/l. Total alkalinity is low for almost all sampled sites except for three sites with high values of alkalinity: one site in Matimba cell (TA = 129.3 mg/l), one in Rwempasha (TA = 73.2 mg/l), and two in Rwimiyaga (TA = 65.9 mg/l and 70.8 mg/l). There is no health-based guideline value for ammonia but according to WHO (2011), natural levels in groundwater and surface water are usually below 0.2 mg/l and anaerobic groundwater may contain up to 3 mg/l. For all sampled sites ammonia concentration is less than 3mg/l except for one site in Rwempasha and one site in Rwimiyaga site (4.6 mg/l and 3.5 mg/l, respectively). NO_2^- and NO_3^- concentrations also respect the WHO (2011) guideline values (2 mg/l and 50 mg/l, respectively). The concentrations of NO_3^- are close to 0 for all sampled sites, confirming that Nyagatare groundwater is anoxic. The concentrations of F^-, Cl^-, PO_4^{3-} and SO_4^{2-} respect the standards. Regarding the concentration of Fe^{2+}, all sampled sites have values exceeding the value of 0.3 mg/l that is the upper acceptable concentration in most national drinking water standards (including Rwanda). Ten sites have values of Mn^{2+} exceeding the value of 0.1 mg/l that is recommended by several national standards to avoid esthetic and operational problems. For other heavy metals, Zn^{2+} respects the norm for all sampled sites except for all Rwempasha and Rwimiyaga sites. Even if the main focus of this research is the removal of heavy metals, the concentrations of Pb^{2+}, Cd^{2+}, Cu^{2+}, As and Cr in Nyagatare groundwater were found to be below the detection limits.

Table 1: Physico-chemical characteristics of Nyagatare groundwater

Sample site	T	pH	Turb.	Cond.	DO	TDS	TA	TH	Ca^{2+}	Na$^+$	K$^+$	Mg^{2+}	NH$_3$	NO$_2^-$
Musheli	27.2	6.3	1.6	355.0	0.01	450.0	46.4	27.0	11.0	3.9	3.0	0.5	0.1	0.07
Matimba	28.3	6.4	3.0	305.0	0.02	428.0	48.8	9.6	4.9	4.6	6.7	0.2	0.2	0.09
Matimba	25.7	6.9	4.0	257.0	0.01	298.0	**129.3**	40.0	14.7	3.0	4.0	8.0	0.5	0.15
Rwempasha	24.9	6.6	5.0	431.0	0.01	492.0	**73.2**	17.6	6.4	4.4	3.6	2.0	0.1	0.12
Rwempasha	27.1	6.4	1.0	283.0	0.01	340.0	46.4	17.4	4.0	6.7	7.8	4.0	0.3	0.10
Rwempasha	27.1	6.6	1.0	457.0	0.02	480.0	39.0	66.0	2.3	4.9	9.1	8.0	**4.6**	0.09
Rwimiyaga	28.3	7.1	1.5	429.0	0.01	489.0	**65.9**	36.0	12.4	5.5	9.0	8.0	**3.5**	0.12
Rwimiyaga	28.0	7.2	1.0	399.0	0.01	490.0	**70.8**	25.0	2.6	2.8	2.0	6.0	0.4	0.14
Rwimiyaga	27.8	7.1	1.0	328.0	0.02	432.0	39.0	29.2	18.0	4.9	7.8	9.0	0.4	0.17
Karangazi	27.2	7.0	2.0	407.0	0.01	487.0	48.8	18.0	9.4	4.9	6.9	5.0	0.1	0.12
Karangazi	26.90	8.2	3.0	301.0	0.01	345.0	13.4	4.2	2.4	8.3	19.4	2.0	0.4	0.13
Karangazi	26.80	9.7	3.0	436.0	0.01	498.0	12.2	10.0	8.0	7.8	18.9	0.1	**1.3**	0.12
Karangazi	26.60	**6.0**	2.0	118.6	0.02	192.0	8.5	28.0	12.0	9.5	13.8	0.2	0.2	0.12
Nyagatare	24.80	6.8	0.0	340.0	0.01	558.0	20.7	18.0	11.1	6.6	23.7	0.8	0.6	0.15
Nyagatare	23.80	**5.4**	0.0	75.0	0.01	79.0	3.4	16.0	9.6	8.9	24.5	0.5	0.1	0.10
Katabagemu	22.90	**6.0**	0.0	234.0	0.02	289.0	8.5	8.6	4.1	4.8	18.0	3.0	0.03	0.12
Katabagemu	22.70	**5.6**	0.0	120.5	0.01	145.0	4.8	32.0	2.9	10.0	15.0	2.0	0.02	0.14
Rukomo	21.50	**5.8**	0.0	375.0	0.01	398.0	6.1	9.6	4.5	11.2	11.0	3.0	0.05	0.12
Tabagwe	21.50	**5.7**	0.0	298.0	0.02	325.0	5.9	7.4	7.2	9.8	24.8	0.1	0.01	0.14
Tabagwe	22.20	**5.6**	1.0	252.0	0.03	275.0	4.8	7.2	5.2	9.9	18.9	1.0	0.01	0.13

Sample sites	NO_3^-	F^-	PO_4^{3-}	Cl^-	HCO_3^-	SO_4^{2-}	Fe^{2+}	Mn^{2+}	Zn^{2+}	Cu^{2+}	Cd^{2+}	Pb^{2+}	As	Cr
Musheli	0.00	0.30	0.37	9.00	28.3	7.60	1.56	**0.97**	0.20	BDE	BDE	BDE	BDE	BDE
Matimba	0.02	1.17	0.15	8.70	29.8	6.00	1.17	**0.60**	0.82	BDE	BDE	BDE	BDE	BDE
Matimba	0.00	0.92	0.30	2.80	78.9	25.00	1.74	**0.20**	0.42	BDE	BDE	BDE	BDE	BDE
Rwempasha	0.03	0.46	0.21	9.90	44.6	5.40	1.22	**0.56**	**6.78**	BDE	BDE	BDE	BDE	BDE
Rwempasha	0.00	0.61	0.20	6.90	28.3	14.00	1.11	**0.35**	**7.09**	BDE	BDE	BDE	BDE	BDE
Rwempasha	0.01	0.34	0.17	12.40	23.8	10.30	1.12	**0.93**	**4.15**	BDE	BDE	BDE	BDE	BDE
Rwimiyaga	0.04	0.77	0.24	7.20	40.2	25.40	1.26	**0.57**	**5.83**	BDE	BDE	BDE	BDE	BDE
Rwimiyaga	0.02	0.57	0.09	9.20	43.2	3.70	1.88	**0.21**	**8.76**	BDE	BDE	BDE	BDE	BDE
Rwimiyaga	0.04	1.13	0.18	18.90	23.8	18.00	1.37	**0.17**	**3.45**	BDE	BDE	BDE	BDE	BDE
Karangazi	0.04	0.50	0.22	8.90	29.8	15.40	1.22	**0.15**	0.91	BDE	BDE	BDE	BDE	BDE
Karangazi	0.02	0.88	0.40	9.80	8.2	19.00	1.11	0.03	0.27	BDE	BDE	BDE	BDE	BDE
Karangazi	0.01	1.07	0.22	5.60	7.4	28.00	1.39	0.03	0.49	BDE	BDE	BDE	BDE	BDE
Karangazi	0.01	0.04	0.21	12.30	5.2	22.00	1.73	0.05	0.56	BDE	BDE	BDE	BDE	BDE
Nyagatare	0.03	0.03	0.24	16.40	12.6	25.00	1.63	0.06	1.38	BDE	BDE	BDE	BDE	BDE
Nyagatare	0.02	0.09	0.16	13.50	2.0	29.00	1.30	0.03	0.58	BDE	BDE	BDE	BDE	BDE
Katabagemu	0.02	0.09	0.31	8.90	5.2	19.00	1.20	0.01	0.00	BDE	BDE	BDE	BDE	BDE
Katabagemu	0.03	0.09	0.22	14.50	2.9	15.00	1.81	0.05	0.63	BDE	BDE	BDE	BDE	BDE
Rukomo	0.04	0.16	0.12	17.90	3.7	9.80	1.22	0.05	0.07	BDE	BDE	BDE	BDE	BDE
Tabagwe	0.02	0.00	0.15	30.40	3.8	9.60	**0.99**	0.01	0.00	BDE	BDE	BDE	BDE	BDE
Tabagwe	0.03	0.46	0.04	22.80	2.9	11.00	1.58	0.01	0.00	BDE	BDE	BDE	BDE	BDE

2.4.2 Piper Tri-linear diagrams

Chemical data of samples from the studied area are presented by plotting them on a Piper-Tri-linear diagram (Figure 2.2). The concept of hydrochemical facies was developed in order to understand and identify the water composition in different classes. Maximum and minimum values of the parameters involved in the Piper diagram are listed in Table 2.2.

Table 2.2: Maximum and minimum concentration of major ions in groundwater samples of Nyagatare

Type of Ion	Minimum concentration (mg/l)	Maximum concentration (mg/l)
Na^+	3.0	11.2
K^+	11.0	24.8
Mg^{2+}	0.1	9.0
Ca^{2+}	2.3	12.0
CO_3^{2-}	0.02	6.9
HCO_3^-	2.1	78.9
Cl^-	12.3	30.4
SO_4^{2-}	10.3	29.0
TDS	145.0	558.0

Before plotting the Piper diagram, the percentage balance error was calculated. For groundwater, the error should be 5% or less unless the total dissolved solids (TDS) value is below 5 mg/l, in which case a higher error is acceptable. If the error exceeds 10%, the analysis should be checked for errors in the transcription or technique (Appelo et al. 1996). The percentage balance error is calculated by using Equation (2.1):

$$\% \ Error \ of \ Ions \ balance = \frac{\Sigma cations - \Sigma anions}{\Sigma cations + \Sigma anions} X100 \qquad (2.1)$$

Table 3: Error of ions balance of groundwater of Nyagatare

Site	1	2	3	4	5	6	7	8	9	10	11	12	13	14	15	16	17	18	19	20
%	4.8	**8.1**	4.7	**6.1**	5.3	**8.6**	4.7	5.3	5.1	2.4	4.4	3.9	2.9	4.5	3.9	4.9	3.1	4.3	5.1	**6.7**

In this study, the percentage balance error varied between 2.1 and 5.1 for 16 samples. The other 4 samples have a percentage balance error more than 5%. This is shown in Table 3. In plotting the Piper diagram, sites with balance error exceeding 5% (sites 2, 4, 6 and 20) were excluded because their analysis was questionable.

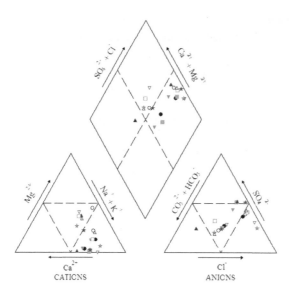

Figure 2.2: Classification of Nyagatare groundwater samples in Piper Tri-linear diagram

Figure 2.2 clearly defines the variations or domination of cation and anion concentrations in different samples. The triangle on the left side presents cation composition, while the triangle on the right side presents the anion composition. The rectangular diagram shows both anions and cations. This plotting procedure assists in the classification of groundwater by its ion ratios. In terms of cations, most of the sampled sites are mainly dominated by sodium and potassium type and for few of them, there is no dominant type. In terms of anions, a few sites have a chloride groundwater type, one has bicarbonate groundwater type and others have no dominant anions.

2.4.3 Principal component analysis

The initial and extraction communalities were obtained from PCA analysis. Initial communalities are estimates of the variance in each variable accounted for, by all components or factors. For extraction of principal components, this is always equal to 1.0 for correlation analyses. Extraction communalities are estimates of the variance in each variable accounted for by the components. The communalities in Table 2.3 are all high, which indicates that the extracted components represent the variables well. If any communality is very low in a principal components extraction, another component may need to be extracted.

Table 2.3: Extraction communalities in Principal Component Analysis

Parameters	Initial	Extraction
T	1.000	.990
pH	1.000	.982
Turbidity	1.000	.972
Conductivity	1.000	.969
DO	1.000	.971
TDS	1.000	.956
TA	1.000	.986
TH	1.000	.969
Ca	1.000	.914
Na	1.000	.847
K	1.000	.965
Mg	1.000	.894
NH_3	1.000	.990
NO_2^-	1.000	.971
NO_3^-	1.000	.960
F^-	1.000	.925
PO_4^{3-}	1.000	.815
Cl^-	1.000	.849
SO_4^{2-}	1.000	.966
Fe^{2+}	1.000	.894
Mn^{2+}	1.000	.935
Zn^{2+}	1.000	.974

The variance explained by the initial solution, extracted components, and rotated components are displayed in Table 2.4. This first section of the table shows the initial eigenvalues. The total column gives the eigenvalues, or amount of variance in the original variables accounted for by each component. The % of variance column gives the ratio, expressed as a percentage, of the variance accounted for by each component to the total variance in all of the variables.

Table 2.4: Initial solution, extracted components, and rotated components during Extraction Method in Principal Component Analysis

Component	Initial Eigenvalues			Extraction Sums of Squared Loadings			Rotation Sums of Squared Loadings		
	Total	% of Variance	Cumulative %	Total	% of Variance	Cumulative %	Total	% of Variance	Cumulative %
1	7.604	34.563	34.563	7.604	34.563	34.563	6.630	30.135	30.135
2	4.474	20.337	54.900	4.474	20.337	54.900	4.046	18.391	48.526
3	3.237	14.712	69.612	3.237	14.712	69.612	3.025	13.752	62.278
4	2.532	11.508	81.120	2.532	11.508	81.120	2.624	11.928	74.206
5	1.716	7.799	88.919	1.716	7.799	88.919	2.332	10.600	84.806
6	1.132	5.147	94.066	1.132	5.147	94.066	2.037	9.260	94.066
7	.731	3.322	97.388						
8	.370	1.680	99.069						
9	.205	.931	100.000						
10	$4.37 \cdot 10^{-16}$	$1.99 \cdot 10^{-15}$	100.000						
11	$3.12 \cdot 10^{-16}$	$1.42 \cdot 10^{-15}$	100.000						
12	$2.34 \cdot 10^{-16}$	$1.06 \cdot 10^{-15}$	100.000						
13	$1.73 \cdot 10^{-16}$	$7.85 \cdot 10^{-16}$	100.000						
14	$1.19 \cdot 10^{-16}$	$5.42 \cdot 10^{-16}$	100.000						
15	$4.87 \cdot 10^{-16}$	$2.21 \cdot 10^{-16}$	100.000						
16	$-6.28 \cdot 10^{-16}$	$-2.86 \cdot 10^{-17}$	100.000						
17	$-9.40 \cdot 10^{-16}$	$-4.27 \cdot 10^{-16}$	100.000						
18	$-1.78 \cdot 10^{-16}$	$-8.09 \cdot 10^{-16}$	100.000						
19	$-1.92 \cdot 10^{-16}$	$-8.75 \cdot 10^{-16}$	100.000						
20	$-2.60 \cdot 10^{-16}$	$-1.18 \cdot 10^{-15}$	100.000						
21	$-3.31 \cdot 10^{-16}$	$-1.51 \cdot 10^{-15}$	100.000						
22	$-6.08 \cdot 10^{-16}$	$-2.76 \cdot 10^{-16}$	100.000						

The cumulative % column gives the percentage of variance accounted for by the first n components. For example, the cumulative percentage for the second component is the sum of the percentage of variance for the first and second components. For the initial solution, there are as many components as variables, and in a correlation analysis, the sum of the eigenvalues equals the number of components. During extraction of the components eigenvalues greater than 1were chosen and the first six principal components are in the extracted solution. The second section of the table shows the extracted components. They explain nearly 94% of the variability in the original 22 variables, so that one can considerably reduce the complexity of the data set by using these components, with only a 6% loss of information. The rotation maintains the cumulative percentage of variation explained by the extracted components, but that variation is now spread more evenly over the components.

The large changes in the individual totals suggest that the rotated component matrix is easier to interpret than the unrotated matrix.

The scree plot presented on Figure 2.4 helps in determining the optimal number of components. The eigenvalue of each component in the initial solution is plotted. Generally, the components to be extracted are on the steep slope, because the components on the shallow slope contribute little to the extracted solution. The last big drop occurs between the sixth and seventh components, so using the first six components is an easy choice.

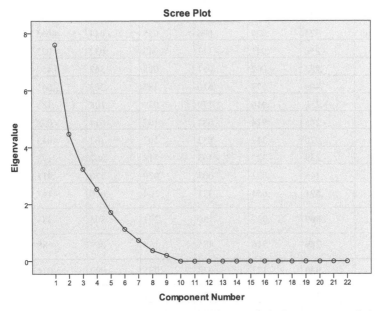

Figure 2.4: Scree plot obtained from the extraction method in the principal component analysis

The rotation matrix was converged in 8 rotations and is presented in Table 2.5. The rotated component matrix helped in determining what the components represent. The first component is most highly correlated with fluoride, pH and sulfate (0.931, 0.922 and 0.914, respectively). The second component is most highly correlated with calcium and total hardness (0.914 and 0.910, respectively) and the third component is most highly correlated with total alkalinity (0.616). The fourth, fifth, and sixth components are mostly correlated with potassium, iron and magnesium, respectively. This suggests that one has to focus on fluoride, pH, sulfate, calcium, total hardness, total alkalinity, potassium, iron and manganese in further analyses. However, ammonia (0.891), turbidity (0.837), nitrite (0.809) and total dissolved solids (TDS) (0.783) also have considerable correlation factors with component 1 and cannot be totally excluded from the discussion on Nyagatare groundwater.

Table 2.5: Rotation method using varimax with Kaiser normalization of Nyagatare groundwater quality

	Component					
	1	2	3	4	5	6
T	.587	-.307	.602	.163	.356	-.190
PH	**.922**	.028	.314	.124	-.041	-.127
Turbidity	**.837**	-.397	.262	-.134	.159	-.047
Conductivity	.525	.824	.096	.055	-.044	-.005
DO	-.250	-.041	-.112	-.943	.032	-.073
TDS	**.783**	-.062	.051	.028	-.362	-.452
TA	.409	.577	**.616**	.187	.263	-.041
TH	-.250	**.910**	.179	-.029	.108	.183
Ca	-.245	**.914**	.081	-.107	-.014	-.032
Na	-.020	-.312	-.852	.102	.103	.043
K	-.239	-.325	-.463	**.718**	.207	.173
Mg	-.145	.092	.004	.070	-.170	**.911**
NH$_3$	**.891**	.054	.351	.204	.040	-.167
NO$_2^-$	**.809**	.039	-.180	.201	.308	.384
NO$_3^-$	-.289	.516	-.485	.158	.069	.588
F$^-$	**.930**	-.177	.078	-.095	-.090	.074
PO$_4^{3-}$	.407	-.217	.579	.227	-.171	.432
Cl$^-$	-.036	.640	-.488	.426	.079	.108
SO$_4^{2-}$	**.914**	.113	-.135	-.033	-.275	-.149
Fe^{2+}	-.119	-.022	-.113	.009	**.923**	-.118
Mn^{2+}	.014	.161	-.017	.670	.672	.088
Zn^{2+}	-.027	.323	.334	.543	.597	-.325

For each case and each component, the component score was computed by multiplying the case's standardized variable values by the component's score coefficients. The resulting six component score variables are representative of, and can be used in place of, the twenty two original variables with only a 6% loss of information. Using the selected components is also preferable to using fluoride, pH, sulfate, calcium, total hardness, total alkalinity, potassium, iron and manganese because the components are representative of all twenty two original variables, and the components are not linearly correlated with each other. Table 2.6 shows the component transformation matrix while Table 2.7 shows the component score coefficient matrix.

Table 2.6: Rotation method using varimax with Kaiser Normalization for Nyagatare groundwater

Component	1	2	3	4	5	6
1	.905	-.111	.371	.073	-.017	-.158
2	.043	.824	.131	.418	.331	.133
3	.040	-.491	-.307	.636	.500	.094
4	.347	.144	-.620	.071	-.454	.513
5	.163	.215	-.598	-.166	.177	-.715
6	.173	-.002	-.094	-.619	.635	.418

Table 2.7: Component score matrix (Varimax with Kaiser Normalization) of Nyagatare groundwater quality

	Component					
	1	2	3	4	5	6
T	.037	-.094	.177	.011	.156	-.003
pH	.129	.022	.022	.029	-.025	-.022
Turbidity	.142	-.088	.015	-.144	.164	.084
Conductivity	.109	.227	-.053	-.027	-.028	-.020
DO	.036	.024	-.068	-.475	.236	.052
TDS	.102	.032	-.084	.099	-.206	-.247
TA	.025	.126	.180	.013	.084	.019
TH	-.027	.214	.071	-.067	.044	.074
Ca	-.026	.236	.014	-.058	-.020	-.058
Na	.085	-.052	-.337	-.005	.076	-.029
K	-.040	-.110	-.114	.312	-.035	.003
Mg	.003	-.024	.094	-.018	-.037	.480
NH_3	.116	.024	.038	.059	-.007	-.046
NO_2^-	.203	.009	-.144	-.091	.212	.247
NO_3^-	.034	.113	-.144	-.021	.047	.245
F^-	.176	-.022	-.061	-.100	.035	.105
PO_4^{3-}	.001	-.097	.256	.113	-.107	.277
Cl^-	.043	.167	-.204	.147	-.046	-.051
SO_4^{2-}	.179	.076	-.168	-.015	-.102	-.077
Fe^{2+}	.036	-.017	-.079	-.206	.499	.005
Mn^{2+}	.000	.002	-.006	.164	.215	.027
Zn^{2+}	-.059	.054	.105	.179	.146	-.183

Although the linear correlation between the components is guaranteed to be 0, one should look at the plots of the component scores (Figure 2.5 and Table 2.8) and check for outliers and nonlinear associations between the components. The first plot in the first row (Figure 3) shows the first component on the vertical axis versus the second component on the horizontal axis, and the order of the remaining plots follows from there. The component score variance matrix is shown in Table 8.

Table 2.8: Component score variance matrix

Componen t	1	2	3	4	5	6
1	1.000	.000	.000	.000	.000	.000
2	.000	1.000	.000	.000	.000	.000
3	.000	.000	1.000	.000	.000	.000
4	.000	.000	.000	1.000	.000	.000
5	.000	.000	.000	.000	1.000	.000
6	.000	.000	.000	.000	.000	1.000

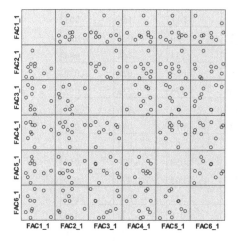

Figure 2.5: Scatter plot matrix from Varimax with Kaiser Normalization

2.4.4 Water hardness, percentage of sodium, residual sodium carbonate and sodium adsorption ratio

Water hardness is caused primarily by the presence of cations such as calcium and magnesium in association with anions such as carbonate, bicarbonate, chloride and sulfate in water. Very hard water is unsuitable for domestic use. In the Nyagatare district, the total hardness varied between 4.2 and 66.2 mgCaCO$_3$/l. According to Sawyer and McCarty's (Sawyer, 1967) classification for hardness, all samples fall under soft class. This suggests that the Nyagatare groundwater is suitable for domestic use.

Wilcox (1995) classified groundwater for irrigation purposes based on percentage of sodium and electrical conductivity. Eaton (1950) as cited by Sadashivaiah (2008) recommended the concentration of residual sodium carbonate to determine the suitability of water for irrigation purposes. The sodium in irrigation waters is usually denoted as percentage of sodium and can be determined using Equation 2 (Wilcox, 1995).

$$\% \, Na^+ = (Na^+ x100)/(Na^+ + K^+ + Ca^{2+} + Mg^{2+}) \tag{2.2}$$

The percentage of Na^+ calculated based on the Equation 2.2 gives values below 20% for 6 samples, and values between 20 and 40% for the 14 remaining samples (Table 9). According to Eaton (1950) and cited by Sadashivaiah (2008), all samples are in good proportion of sodium so that they can be used for irrigation.

In waters having high concentrations of bicarbonate, there is a tendency for calcium and magnesium to precipitate as the water in the soil becomes more concentrated. As a result, the relative proportion of sodium in the water is increased in the form of sodium carbonate. The residual sodium carbonate (RSC) is calculated using Equation 2.3.

$$RSC = (HCO_3^- + CO_3^{2-}) - (Ca^{2+} + Mg^{2+}) \tag{2.3}$$

The calculated RSC values of all samples are below 2.5 as shown in Table 2.9, which indicates that the groundwater of Nyagatare is suitable for irrigation.

The sodium adsorption ratio (SAR) was also calculated in order to check of the suitability of the Nyagatare groundwater for use in agricultural irrigation. It was calculated from the concentrations of solids dissolved in the water according to the Equation 2.4 (Mohsen et al. 2009).

$$SAR = \frac{[Na^+]}{\sqrt{(\frac{[Ca^{2+}]+[Mg^{2+}]}{2})}} \tag{2.4}$$

Where Na^+, Ca^{2+} and Mg^{2+} are expressed in meq/L and SAR expressed in $(meq/L)^{1/2}$. The threshold for SAR is 12 $(meq/L)^{1/2}$ and the higher the sodium adsorption ratio, the less suitable the water is for irrigation. Irrigation using water with high sodium adsorption ratios may require soil amendments to prevent long-term damage to the soil (Mohsen et al. 2009). Thus, based on the values of SAR (Table 2.9), Nyagatare groundwater is suitable for irrigation because SAR values for all sampled sites are below the threshold value.

Table 2.9: Sodium adsorption ratio for groundwater of Nyagatare

Site	1	2	3	4	5	6	7	8	9	10	11	12	13	14	15	16	17	18	19	20
% Na^+	21	28	10	27	30	20	16	21	12	19	26	22	27	16	21	16	33	38	23	28
RCS	2.1	0.6	0.8	0.9	1.2	1.4	2	0.8	1.9	1.2	0.9	1.1	0.7	1.7	1.4	2.1	2.0	1.1	1.8	2.3
SAR	1.6	2.9	0.9	2.1	3.4	2.2	1.7	1.4	1.3	1.8	5.6	3.9	3.8	2.7	4.0	2.5	6.4	5.8	5.1	5.6

2.4.5 Discussion of other parameters

Other parameters showed by the PCA as correlating with component 1 are below commented. These are pH, ammonia, iron and manganese, fluoride, Electrical conductivity (EC) and Total dissolved solids (TDS) and ammonia and nitrate.

Iron and Manganese

Iron (Fe) and manganese (Mn) are metals that occur naturally in soils, rocks and minerals. In the aquifer, groundwater comes in contact with these solid materials dissolving them, releasing their constituents, including Fe and Mn, to the water. At concentrations approaching 0.3 mg/L Fe and 0.1 mg/L Mn, the water's usefulness may become seriously impacted, e.g., there may be a metallic taste to the water and staining of plumbing fixtures may become common. At these concentrations, however, the health risk of dissolved Fe and Mn in drinking water is insignificant. If the groundwater is oxygen poor, iron (and manganese) will dissolve more readily, particularly if the pH of the water is on the low side (slightly more acidic). The iron content was measured in the water samples as Fe^{2+}, which is the reduced state of iron. The concentration of iron, in groundwater from 20 wells included in this study, varied between 1.0 and 1.9 mg/L and that of manganese between 0.01 and 0.97 mg/l. The presence of Fe^{2+} in the water samples reveals that the aquifer is anoxic in some zones. This is not surprising as natural water bodies are generally stratified, with an oxic layer lying above a middle zone of progressively decreasing dissolved oxygen content and a totally anoxic layer at the bottom. In comparison with the old WHO guidelines (WHO 1996) for drinking water (0.3mg/L for Fe and 0.1 mg/l for Mn), the groundwater of Nyagatare exceeds the norms for most of the wells analysed.

Fluoride

Fluoride concentrations in Nyagatare groundwater varied between 0.1 and 1.17 mg/l. Waters with a high fluoride content are found mostly in calcium-deficient groundwater in many aquifers, such as granite and gneiss, in geothermal waters and in some sedimentary basins. Fluoride is a desirable substance: it can prevent or reduce dental decay and strengthen bones, thus preventing bone fractures in older people. Where the fluoride level is naturally low, studies have shown higher levels of both dental caries (tooth decay) and fractures (Luke, 1997 and 2001, NRC 2006). Fluoride levels above 1.5 mg/l may, however, have long-term undesirable effects (Fluorosis). Calcium fluorite is a common fluoride mineral which occurs in igneous and sedimentary rock. Since the Nyagatare District has abundant granite and granite rocks being igneous rocks, this can explain the source of fluoride found in the groundwater.

Electrical conductivity (EC) and total dissolved solids (TDS)

The salinity is a measure of the amount of dissolved salts and ions in water. There are several different ways to measure salinity; the two most frequently used analyses being (1) Total Dissolved Solids (TDS): a measure of all dissolved substances in water, including organic and suspended particles that can pass through a 0.45 µm filter. (2) Electrical Conductivity (EC): The ability of an electric current to pass through water is proportional to the amount of dissolved salts in the water. EC values for Nyagatare groundwater varied between 75 and 457 µS/cm while the TDS varied between 79 and 498 mg/l. Such values are quite good for groundwater. The source of EC and TDS and mineral in Nyagatare groundwater can be related to the natural background of soil type but also to the human activities by application of fertilizers and manures.

Ammonia and nitrite
High concentrations of ammonia in groundwater, while not directly harmful to human health, is often a sign of the groundwater being affected by anthropogenic activities, such as spreading of fertilizers and manure or infiltrating sewage water. Nyagatare groundwater showed values of ammonia ranging between 0.1 and 4.6 mg/L. However, these values of ammonia are in the range of drinking water norms. The nitrite concentrations in Nyagatare groundwater range between 0.07 and 0.17 mg/l. All the values of nitrite are within the guideline values of 3 and 0.2 mg/l for short and long term exposure, respectively, as recommended by WHO (2011). The source of ammonia and nitrite is related to the animal manure because the livestock is the main activity in the area.

2.5 Conclusions

In this study, groundwater samples were collected from 20 boreholes in the Nyagatare District and 22 parameters were analysed. The following conclusions are established:
- Most of sampled sites have the values of the analyzed parameters respecting the guidelines for drinking water.
- The Nyagatare groundwater shows elevated content of Fe^{2+} and Mn^{2+} and other heavy metals were not detected.
- Presence of iron as Fe^{2+}, Mn^{2+}, NH_3, and NO_2^- suggests that the Nyagatare groundwater is anoxic
- The Piper diagram showed that most of groundwater from sampled sites is mainly sodium and potassium type and, for a few of wells, there is no dominant groundwater type. In terms of anions, few sites have chloride groundwater type, one has bicarbonate groundwater type and others have no dominant anions.
- PCA results showed that the 6 extracted components represent the variables well explaining nearly 94% of the variability in the original 22 variables. The first component was most highly correlated with fluoride, pH and sulfate, the second component was most highly correlated with calcium and total hardness and the third component is most highly correlated with total alkalinity. The fourth the fifth and the sixth components are mostly correlated with potassium, iron and magnesium, respectively.
- Regarding the hardness of the water, the Nyagatare groundwater is suitable for domestic use except for 3 sites (Rwempasha, Rwimiyaga and Matimba sites)
- The calculation of %Na^+, RSC and SAR showed that Nyagatare groundwater also is suitable for irrigation.
- Nyagatare District having abundant granite and granite rocks being igneous rocks, this can explain the source of fluoride found in groundwater. The source of EC, TDS,

ammonia and nitrite in Nyagatare groundwater can be related to human activities by application of fertilizers and manures.

2.6 References

American Public Health Association (APHA) 1994, Standard method for examination of water and wastewater, *NW, DC 20036*.

AQUASTAT (2005), Systèmes d'information de la FAO sur l'eau et l'agriculture/. Food and Agriculture Organization of the United Nations (FAO), Rome.

Baligira R. (2007), Examen critique et enquête de l'état de la gestion des données de la qualité de l'eau au Rwanda ainsi que les recommandations pour son amélioration., Nile Transboundary Environmental Action Project, Nile Basin Initiative (NBI), Kigali.

Jae-On Kim, Charles W. Mueller (1978) Factor Analysis: Statistical Methods and Practical Volume 14;

Kabalisa V. P., (2006). Analyse contextuelle en matière de Gestion Intégrée des Ressources en Eau au Rwanda, Document de Travail pour l'ONG Protos (Rapport final)

Luke J. (1997). The Effect of Fluoride on the Physiology of the Pineal Gland. Ph.D. Thesis. University of Surrey, Guildord.

Luke J. (2001). Fluoride deposition in the aged human pineal gland. *Caries Research* 35: 125- 128.

MINITERE-ISAR (2007) Rapport d'inventaire des ressources ligneuses au Rwanda, Volume 2, Ministère des Terres, de l'Environnement, des Forêts, de l'Eau et des Mines (MINITERE)-National Agricultural Research Institute (ISAR), Kigali.

Mohsen S.,Majid R. and Borzoo G. K.(2009), Prediction of Soil Exchangeable Sodium Percentage Based on Soil Sodium Adsorption Ratio, American-Eurasian J. Agric. & Environ. Sci., 5 (1): 01-04

NRC (2006). National Research Council of the National Academies, Fluoride in Drinking Water: A Scientific Review of EPA's Standards. Washington, DC: National Academies Press.

Osodo P. and Rwamugema H. (2001), UNICEF-GOR, Water and environmental sanitation programme

Quirk, J.P., (2001), The significance of the threshold and turbidity concentrations in relation to sodicity and microstructure. Australian J. Soil Res., 39: 1185-1217

Sadashivaiah C., Ramakrishnaiah C. R. and Ranganna G. (2008), Hydrochemical Analysis and Evaluation of Groundwater Quality in Tumkur Taluk, Karnataka State, India, *Int. J. Environ. Res. Public Health, 5(3)* 158-164

Sawyer G. N.; McCarthy D. L. (1967), Chemistry of sanitary Engineers, *2nd ed, McGraw Hill, New York*, p-518

Schroeder. H. A. (1960): Relations between hardness of water and death rates from certain chronic and degenerative diseases in the United States, *J. Chron disease, 12*:586-591

Stevens J.P., (1996), Applied Multivariate Statistics for the Social Sciences, Fifth Edition

WHO (1996) Guidelines for drinking water quality, 2nd Edition, Volume 2, Health criteria and other supporting information, Geneva

WHO (2004) Guidelines for drinking water quality, 3rd Edition, Volume 2, Geneva, pp 186-398

WHO (2011), Guidelines for Drinking water quality,: Health Criteria and other supporting information. *World health organization*, Geneva, Switzerland Second edition Vol. 2

Wicox, L. V. (1995), Classification and use of irrigation waters, US Department of Agriculture, *Washington Dc*, 1995, p-19

Zaporozee, A. (1972), Graphical interpretation of water quality data, *groundwater*, 10, 32-43.

Chapter 3: Effect of Calcium on adsorptive removal of As(III) and As(V) by iron oxide based adsorbents

Part of this chapter has been presented as:
V. Uwamariya, B. Petrusevski, P. Lens, G. Amy (2012), Effect of water matrix on adsorptive removal of heavy metals from groundwater. In proceeding of the *IWA World Water Congress and Exhibition*, Busan (Korea) 16-21 September 2012

V. Uwamariya, B. Petrusevski, P. N.L. Lens and G. Amy (2013), Effect of Calcium on Adsorptive Removal of Arsenic from Groundwater by Iron Oxide Based Adsorbents, *Journal of Environmental Technology*, submitted

Abstract

The effects of calcium on the equilibrium adsorption capacity of As(III) and As(V) onto iron oxide coated sand (IOCS) and granular ferric hydroxide (GFH) were investigated through batch experiments, rapid small scale column tests (RSSCT) and kinetics modeling. Batch experiments showed that at calcium concentrations $\leq$ 20 mg/l, higher As(III) and As(V) removal efficiencies by IOCS and GFH were observed at pH 6. An increase of the calcium concentration to 40 and 80 mg/l reversed this trend giving higher removal efficiency at higher pH (8). The adsorption capacities of IOCS and GFH at an equilibrium arsenic concentration of 10µg/l were found to be between 2.0 and 3.1 mg/g for synthetic water without calcium and between 2.8 and 5.3 mg/g when 80 mg/l of calcium was present at all studied pH values. After 10 hours of filter run in RSSCT, and for approximately 1000 Empty Bed Volumes (EBV), the ratios of C/C_o for As(V) were 26% and 18% for calcium-free model water; and only 1% and 0.2% after addition of 80 mg/l of Ca for filter columns with IOCS and GFH, respectively. The adsorption of As(III) and As(V) onto GFH follows a second order reaction with and without addition of calcium while the adsorption of As(III) and As(V) onto IOCS follows a first-order reaction without calcium addition, and moves to the second reaction order kinetics when calcium is added. Based on the intraparticle diffusion model, the main controlling mechanism for As(III) adsorption is intraparticle diffusion, while the surface diffusion contributes greatly to the adsorption of As(V).

Key words: Adsorption, Arsenic, Calcium, kinetics, GFH, IOCS,

3.1. Introduction

The presence of arsenic in groundwater used for drinking water supply is reported in many parts of the world and is a global issue. Long-term exposure to low arsenic concentrations in drinking water is associated with human carcinogen risks, like skin and several internal cancers (Smith *et al.* 2000). The World Health Organization (WHO 2004) recommends a concentration of 10 µg/l as guideline value for arsenic in drinking water. As a naturally occurring element in the earth's crust, arsenic enters into aquifers through natural processes like mineral dissolution (e.g., pyrite oxidation), and reductive desorption and dissolution (Smedley and Kinniburgh, 2002). Arsenic occurs in groundwater predominantly in inorganic forms, with speciation and valence depending on the oxidation-reduction conditions in the aquifer and the pH of the water. Generally, the reduced form of arsenic, As(III) - arsenite, is found in groundwater under anoxic conditions and the oxidized form, As(V)-arsenate, is generally found under oxic conditions. However, both forms can be found in the same drinking water source. Depending on the pH, arsenate exists in four forms in aqueous solution (H_3AsO_4, $H_2AsO_4^-$, $HAsO_4^{2-}$, and AsO_4^{3-}) and arsenite exists in five species ($H_4AsO_3^+$, H_3AsO_3, $H_2AsO_3^-$, $HAsO_3^{2-}$, and AsO_3^{3-}) (Wang *et al.* 2000). Over the pH range common for groundwater (6.5-8.5), predominant As(V) species are $H_2AsO_4^-$ and $HAsO_4^{2-}$, while As(III) is present as the neutral species H_3AsO_3 (Wang *et al.* 2000).

Different technologies have been developed and applied for arsenic removal. Large-scale treatment facilities often use conventional coagulation with alum or iron salts, followed by filtration (Scott et al. 1995, Hering *et al.* 1996, Chen *et al.*, 1999). The lime softening and iron removal by aeration-filtration are also common conventional treatment processes that can partially remove arsenic from groundwater (Mcneill and Edwards 1997). Recently, iron oxide-based adsorption media, such as iron oxide coated sand (IOCS), and some commercial media like Aqua-Bind MP, ArsenX, Bayoxide E33 ferric oxide and Granular Ferric Hydroxide (GFH) have been developed and have demonstrated high arsenic removal capacities in laboratory and pilot tests. However, their full-scale applications are still limited (Petrusevski *et al.* 2002, 2007, Amy *et al.* 2005).

Removal of arsenic by adsorption on iron oxide surfaces results in the formation of a surface complex between soluble arsenic and a solid hydroxide surface (Jeong et al. 2007). Iron oxide coating on the surface of sand increases the positive charge of the surface. Additionally, porosity increases, which further results in a higher adsorption capacity (Benjamin et al. 1996). Several hypothetical models have been proposed to describe the mechanism of arsenic removal by IOCS. Table 3.1 describes the general process of arsenic adsorption on an iron oxide coated surface (Dzombak and Morel 1990).

Table 3.1: Arsenic adsorption on an iron oxide coated surface

Arsenite adsorption	*Arsenate adsorption*
$FeOH + H_3AsO_3 \rightarrow Fe\text{-}H_2AsO_3 + H_2O$	$FeOH + AsO_4^{3-} + 3H^+ \rightarrow Fe\text{-}H_2AsO_4 + H_2O$
	$FeOH + AsO_4^{3-} + 2H^+ \rightarrow Fe\text{-}HAsO_4^- + H_2O$
	$FeOH + AsO_4^{3-} + H^+ \rightarrow Fe\text{-}AsO_4^{2-} + H_2O$

Goldberg and Johnston (2001) have studied the mechanisms of arsenic adsorption on amorphous oxides using macroscopic measurements, vibrational spectroscopy, and surface complexation modeling. Their results showed that arsenate forms inner-sphere surface complexes on both amorphous Al and Fe oxides, whereas arsenite forms both inner-and outer-sphere surface complexes on amorphous Fe oxide and outer-sphere surface complexes on amorphous Al oxide. Surface complexation of As(V) to iron(III) (hydr)oxides has also been studied by Sherman and Randall (2003). They found that adsorption of arsenate onto goethite, lepidocrocite, hematite and ferrihydrite occurs by the formation of inner-sphere surface complexes resulting from bidentate corner-sharing between the AsO_4^{3-} tetrahedral and the FeO_6 polyhedra. They did not find evidence for monodentate complexes of As sorbed to any of the iron oxides or oxide hydroxides. The arsenate retention mechanism on goethite has also been studied by Scott and Morgan (1995) using Extended X-ray Absorption Fine Structure (EXAFS) spectroscopy. They concluded that monodentate, bidentate-binuclear, and bidentate-mononuclear complexes are formed on the surface of goethite. Kundu and Gupta (2006) proposed a mechanism involved in the removal of As(III) by iron oxide coated cement over acidic pH range based on the assumption that the alkaline adsorbent surface helps in the conversion of non-ionic As(III) to its anionic form, which in turn assists in the adsorption process.

Arsenic removal research is often carried out using synthetic water in single ion systems, while, in natural groundwater; arsenic is always accompanied by other multivalent anions. In fact, raw groundwater contains a mixture of many ions which may enhance adsorption, act relatively independently, or compete with one another. For instance the presence of phosphate, sulphate, carbonate, silica, and other anions has been shown to decrease adsorption of arsenic to varying degrees depending on their concentrations, pH, and arsenic speciation (Mohan and Pittman 2007). Experiments on the effect of silica and pH on As(V) adsorption by a resin/iron oxide hybrid media (Möller and Sylvester 2008) showed that the adsorption of As(V) decreases with increasing silica concentration and pH of the solution. Similar batch experiments on the effect of silica, calcium, and pH on As(V) sorption to oxide surfaces (amorphous iron hydroxide and activated alumina) over a pH range between 7 and 12 showed that the adsorption of As(V) significantly decreased in the presence of silica, while the presence of silica together with calcium slightly increased the adsorption of arsenic on the media (Smith and Edwards 2005).

The main objective of the present study was to determine whether calcium, which is a cation commonly found in groundwater, has an effect on the removal efficiency of As(III), and As(V), by IOCS and GFH at different pH values. Experimental conditions were varied to identify the effect of pH and increase in calcium concentration on the removal kinetics and capacity.

3.2 Materials and Experimental Methods

3.2.1 Adsorbents

GFH was obtained from the manufacturer GEH Wasserchemie in Osnabruck (Germany). IOCS was obtained from the water treatment plant Brucht (Dutch Water Supply a Company Vitens) which treats groundwater with a high iron content. For the screening and adsorption isotherm experiments, pulverized IOCS and GFH ($< 63\mu m$) were used. For RSSCT, a 356-390 μm size fraction of IOCS and GFH was used. IOCS was ground with a mortal and a pestle which was followed by sieving to obtain the required size fraction. Grinding was done with care as to obtain only the coated part of IOCS (excluding the sand core itself) for both batch and RSSCT experiments. Table 3.2 summarizes the physical properties of GFH and IOCS.

Table 3.2: Physical characteristics of GFH and IOCS

	d_{60} (mm)	Surface area (m^2/g)	pH$_{ZPC}$	Porosity (%)	Bulk density (g/cm^3)
GFH	1.12	555	7.5-8.0	0.75-0.80	1.32
IOCS	3.58	261	6.9	0.53	1.05

The model water used in this study was prepared by mixing demineralised water with the required amount of As(III) or As(V) from a stock solution of 1000 mg/L of As(V) or As(III). 0.3 mmol of $NaHCO_3$ was added to increase buffering capacity. Subsequently, varying amounts of calcium were introduced as $CaCl_2$.

3.2.2 Batch adsorption experiments

Several batch adsorption experiments were carried out to investigate the effect of calcium on arsenic removal. Short screening adsorption batch experiments of 24 hours were conducted with different calcium concentrations (0, 20, 40 and 80 mg/l) at pH 6, 7 and 8. Long batch experiments with a contact time of 30 days were also carried out to establish adsorption isotherms for As(III) and As(V) adsorption on IOCS and GFH in the presence and/or absence of calcium. For the above mentioned experiments, acid-cleaned and closed 500 ml plastic bottles, fitted with tubes for periodic sampling, were filled with synthetic water and 0.5g/l pulverized IOCS or GFH was added. Subsequently the pH was adjusted to the required values using 1M HNO_3 or NaOH solutions. Bottles were placed on an Innova 2100 rotary shaker at 100 rpm and kept at 20±1°C. Blank tests were carried out without the adsorbent addition. All samples were filtered through a 0.45 μm membrane filter using a polypropylene syringe filter.

3.2.3 RSSCT experiments

In addition to batch adsorption experiments, RSSCT experiments were conducted assuming constant diffusivity (X=0), as shown by Equations (3.1) and (3.2)

$$\frac{EBCT_{SC}}{EBCT_{LC}} = \left[\frac{d_{psc}}{d_{plc}}\right]^{2-X} = \frac{t_{sc}}{t_{lc}} \tag{3.1}$$

$$\frac{V_{SC}}{V_{LC}} = \left[\frac{d_{pLc}}{d_{pSc}}\right] \tag{3.2}$$

where *EBCT*= empty bed contact time; *SC* = short column; *LC* = large column; d_p= IOCS/GFH grain diameter; t = filter run time; *V*= loading rate (Westerhoff *et al.*, 2005)

Glass columns with an inner diameter of 1.8 cm were packed with IOCS or GFH. After filling the columns with IOCS or GFH, the set-up was operated with demineralised water to rinse the media and to release air from the filter bed. The flow direction was downward. Samples of feed water and filtrate were taken at regular time intervals of 1 hour during filter runs of 10 hours. Columns were operated at filtration rates of 46.6 m/h and an empty bed contact time (EBCT) of 0.35 min.

3.2.4 Analytical techniques

Arsenic was analyzed with an atomic absorption spectrometer (Thermo Elemental Solaar MQZe-GF 95) with an auto-sampler and a graphite furnace used as detector (AAS-GF).

Acidified demineralised water was used for dilution. Nickel nitrate (50 g Ni/l) was used as matrix modifier. IRA-400 (chloride form) anion resin was used to separate As (III) and As (V). Water samples were passed through a 45μm filter and acidified with HCl to a pH below 2. Acidified samples were passed through the anion exchange resin column which only allows arsenite passage. Arsenate was determined as the difference between total arsenic and arsenite (Bissen et al. 2000). Calcium was analyzed with an atomic absorption spectrophotometer-Flame (Perkin-Elmer model AAnalyst 200).

3.2.5 Models applied

PHREEQC modelling

The PHREEQC-2 hydro geochemical model was used to calculate the saturation indices of different species formed in model water with different concentrations of calcium and different pH values. In this model the water composition is input to the model and PHREEQC calculates the ion activities and saturation states for relevant minerals (Appelo and Postma, 2005).

Isotherm models

The experimental results from long batch experiments were fitted in the Freundlich isotherm model (Equation 3.3):

$$q = K\, Ce^{1/n} \tag{3.3}$$

Where q is the amount of solute adsorbed per unit weight of adsorbent (mg/g), Ce represents the equilibrium concentration of the solute (mg/l), and K and $1/n$ are the isotherm constants also known as the fitting parameters. K [(mg/g) x (mg/L)^n] is the measure of adsorption capacity and $1/n$ is the measure of adsorption intensity. Keeping the value of $1/n$ fixed, the larger the value of K, the greater the adsorption capacity. For a fixed value of K, the smaller the value of $1/n$ the more favorable the adsorption (Massel, 1996).

Adsorption kinetics models

Adsorption kinetics models have been widely used to describe the kinetic process of adsorption (Chen *et al.*, 2008; Hameed, 2008; Huang *et al.*, 2008; Wan Ngah and Hanafiah, 2008; Rosa *et al.*, 2008; Tan *et al.*, 2008).

There are several models available that describe adsorption kinetics. In the present study, pseudo first order, pseudo-second order, intraparticle diffusion models and the Elovich kinetic equations were applied.

Pseudo-first order kinetic model

The first order rate expression is based on the Lagergren Equation. It describes the kinetic process of liquid-solid phase adsorption of oxalic acid and malonic acid onto charcoal (Lagergren, 1989). It is represented by the Equation 3.4 as:

$$\frac{dq}{dt} = k_1(q_e - q_t) \tag{3.4}$$

Where q_t and q_e (mg/g) are the amount of solute adsorbed at time t (min) and at equilibrium, respectively. k_1 (min^{-1}) is the pseudo-first order rate constant for the kinetic model. By integrating Equation (3.4) with the boundary conditions of $q_t = 0$ at $t = 0$ and $q_t = q_t$ at $t = t$, yields (Ho, 2004):

$$\log(q_e - q_t) = \log(q_e) - \frac{k_1}{2.3030}t \qquad (3.5)$$

The straight line obtained by plotting log (q_e- q_t) against t gives log q_e as the intercept and - $k_1/2.303$ as the slope. Thus, the amount of solute adsorbed at equilibrium (q_e) and the first order kinetic constant (k_1) can be obtained from the intercept and the slope.

Pseudo-second order kinetic model

The pseudo-second order rate equation is described by the Equation 3.6. The driving force, (q_e-q_t), is proportional to the available fraction of active sites (Ho, 2006).

$$\frac{dq_t}{dt} = k_2(q_e - q_t)^2 \qquad (3.6)$$

Integrating Equation (3.6) with the boundary conditions of $q_t = 0$ at $t = 0$ and $q_t = q_t$ at $t = t$, yields

$$\frac{1}{(q_e - q_t)} = \frac{1}{q_e} + k_2 t \qquad (3.7)$$

Where k_2 is the second order rate constant. Equation (3.7) can be rearranged in order to obtain a linear equation:

$$\frac{t}{q_t} = \frac{1}{k_2 q_e^2} + \frac{1}{q_e}t \qquad (3.8)$$

The initial adsorption rate, h, is defined as:

$$h = k_2 q_e^2 \qquad (3.9)$$

The plot of t/qt against t gives a straight line with the intercept equals to $1/k_2 q_e^2$ and a slope of $1/q_e$. Hence the amount of solute adsorbed per gram of sorbent at equilibrium (q_e) and the sorption rate constant (k_2) are evaluated from the intercept and the slope, respectively.

Intraparticle diffusion model

Adsorption normally occurs through 3 different steps: (1) external diffusion or film diffusion; (2) internal diffusion or intra-particle diffusion; and (3) adsorption and desorption between the adsorbate and active sites, i.e., mass action. The second step was assumed to be the rate controlling step was used. The intraparticle diffusion model is given by the Equation (3.10) (Acheampong et al., 2012):

$$q_t = k_{id} t^{1/2}$$

$$(3.10)$$

where k_{id} is the intraparticle diffusion rate constant ($mg.g^{-1}min^{-1/2}$). The plot of q_t vs $t^{1/2}$ represents different stages of adsorption; the linear portion of the plot q_t versus $t^{1/2}$ represents the intraparticle diffusion. The slope yields the intraparticle diffusion rate constant K_{id} and the intercept reflects the boundary layer effect. The bigger the intercept, the greater is the contribution of the surface sorption to the rate-controlling step.

The Elovich equation

The Elovich equation is a kinetic equation of chemisorption. It was established by Zeldowitsch in 1934 and was used to describe the adsorption rate of carbon monoxide on manganese dioxide that decreases exponentially with an increase in the amount of gas adsorbed (Ho, 2006). The Elovich equation is expressed as follows:

$$\frac{dq_t}{dt} = ae^{-\beta q_t}$$

$$(3.11)$$

where q_t represents the amount of gas adsorbed at time t, β the desorption constant, and α the initial adsorption rate (Ho and McKay, 2000). Equation (3.11) can be rearranged to a linear form:

$$q_t = \left[\frac{2.3}{\alpha}\right]\log(t + t_o) - \left[\frac{2.3}{\alpha}\right]\log t_o$$

$$(3.12)$$

Where $t_o = \frac{1}{\alpha}\beta$

Assuming that $\alpha\beta t >> 1$, integrating and applying boundary conditions $qt = 0$ at $t = 0$ and $q_t = qt$ at $t = t$, Equation (3.12) becomes:

$$q_t = \frac{1}{\beta}\ln(\alpha\beta) + \frac{1}{\beta}\ln(t)$$

$$(3.13)$$

A plot of q_t versus $ln(t)$ gives $1/\beta$ as the slope and $1/\beta ln(\alpha\beta)$ as the intercept.

3.3. Results

3.3.1 PHREEQC-2 modeling

At pH 7, the species of As(III) predicted to be in the solution are H_3AsO_3, $H_2AsO_3^-$, $H_4AsO_3^+$, $HAsO_3^{-2}$ and AsO_3^{-3}. The predominant form is neutral H_3AsO_3 representing 99.6% of the total concentration of As(III) in the solution. Mineral forms of As(III) predicted in the solution are arsenolite (As_2O_3) and claudetite (As_4O_6) with negative saturation indices of -14.21 and -13.92, respectively.

The As(V) species predicted in the solution are H_3AsO_4, $H_2AsO_4^-$, $HAsO_4^{-2}$, AsO_4^{-3}, H_3AsO_4. The predominant form of As(V) are $HAsO4^{-2}$ and $H_2AsO_4^-$ with 55.9 and 43.9%, respectively, of the total concentration of As(V) in the solution. The predicted form of As(V) to precipitate in the solution was arsenic pentoxide (As_2O_5) with a negative saturation index of -25.67.

3.3.2 Short term adsorption batch experiments

Short term batch experiments were carried out to screen the effect of calcium on the removal of both As(III) and As(V) by IOCS and GFH. Calcium concentrations were varied from 0 to 80 mg/l. The results obtained are presented in Figures 3.1 and 3.2.

Figure 1A shows that the presence of calcium in synthetic water enhanced As(III) adsorption on IOCS at all pH values investigated. The beneficial effect was most pronounced at pH 8 with an increase of the As(III) removal efficiency from about 40% (without calcium) to more than 80% (with 80 mg/l of calcium). At calcium concentrations ≤ 20 mg/l, higher As(III) removal efficiencies were observed at lower pH values. An increase of calcium concentration from zero to 40 and 80 mg/l reversed this trend giving higher removal efficiencies at higher pH values. In fact, at pH 8 and calcium concentration ≤ 20 mg/l, As(III) was removed at 40% while at calcium concentrations of 40 and 80 mg/l, the removal efficiencies of As(III) were 82 and 87%, respectively, at pH 8. The effect of calcium on As(V) removal by IOCS (Figure 1B) was limited at pH values of 6 and 7, while it strongly improved As(V) removal at pH 8, from 48% removal efficiency in the model water without calcium to 79 % when 80 mg/l of calcium was present in the model water. As(V) was better removed than As(III) by the IOCS. For both forms of arsenic, the removal was better when calcium was present.

[A]

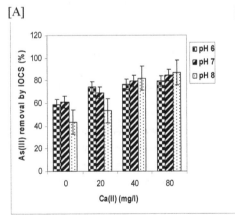

[C]

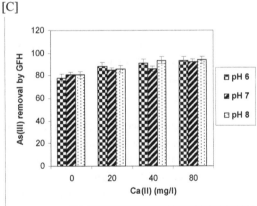

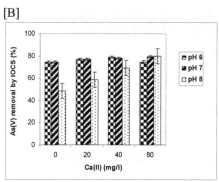
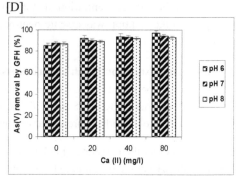

Figure 3.1: Effect of calcium on As(III) (A) and As(V) (B) removal by IOCS and effect of calcium on As(III) (C) and As(V) (D) removal by GFH at different pH values. Model water composition: initial As concentration = 3880 µg/l, HCO_3^- = 20 mg/l, adsorbent dosage = 0.5 g/l, contact time = 24hours

The presence of calcium in the model water had a similar effect on As(III) and As(V) adsorption by GFH. As(V) removal by GFH was, in general, somewhat better than As(III), as shown in Figure 3.1C and 3.1D. Removal of As(III) by GFH increased slightly with increasing calcium concentration from 0 to 80 mg/l. At pH 6, the removal of As(III) varied from 78% to 93%, while it varied from 80% to 92% at pH 7 and pH 8 (Figure 3.1C). A similar effect of the presence of calcium in model water was observed for As(V) adsorption on GFH (Figure 3.1D), with a slight increase in the removal efficiencies at all pH values when calcium was present (between 6 and 12% increase as a function of model water pH). The pH had only a very limited effect on the As(III) removal efficiency by IOCS/GFH.

3.3.3 Batch isotherm experiments

Thirty day batch experiments for As(III) and As(V) adsorption on IOCS and GFH were conducted at pH 7 to establish the isotherms in the presence and absence of calcium. The results obtained fit the Freundlich model (Figure 3.2), with high correlation coefficients (R^2 = 0.91-0.97) (Table 3.2).

The Freundlich isotherm constants (K) obtained ranged between 845 and 3357 (mg/g) $(L/mg)^{1/n}$. The highest adsorption capacity was observed for adsorption of As(V) on GFH in the presence of calcium, and the lowest being the adsorption of As(III) on IOCS in the absence of calcium. The coefficient 1/n exhibited relatively low values varying between 0.20-0.38. Freundlich isotherms are shown on Figure 3.2. The adsorption capacities of IOCS and GFH calculated for an equilibrium arsenic concentration of 10µg/l (Table 3.3) were found to be between 2.0 and 3.1 mg/g for synthetic water without calcium and between 2.8 and 5.3 mg/g when 80 mg/l of calcium was present. Results from long batch isotherm experiments of 30 days confirmed the beneficial effect of calcium on As(V) and As(III) removal as obtained in short batch adsorption tests of 24 hours

A   B

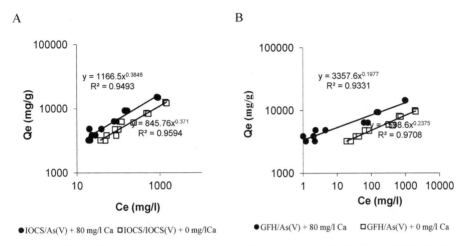

● IOCS/As(V) + 80 mg/l Ca   □ IOCS/IOCS(V) + 0 mg/lCa   ● GFH/As(V) + 80 mg/l Ca   □ GFH/As(V) + 0 mg/l Ca

Figure 3.2: Freundlich adsorption isotherms for As(III) adsorption on IOCS (A) and GFH (B). Model water [As] = 3.880 mg/l, $[Ca^{2+}]$ = 0 and 80 mg/l, HCO_3^- = 20 mg/l, pH 7

Table 3.3: Freundlich isotherms parameters

EXPERIMENT		K [(mg/g).(mg/L)^n]	1/n	R^2	Q for Ce = 10 µg/l (mg/g)
As(III) removal by IOCS	Without Ca	850	0.371	0.959	1.98
	With Ca	1170	0.385	0.949	2.83
As(V) removal by IOCS	Without Ca	1610	0.211	0.924	2.61
	With Ca	2110	0.265	0.923	3.88
As(III) removal by GFH	Without Ca	1600	0.238	0.971	2.76
	With Ca	2550	0.259	0.927	4.63
As(V) removal by GFH	Without Ca	1860	0.219	0.913	3.07
	With Ca	3360	0.198	0.933	5.29

3.3.4 Adsorption kinetic study

The values of the kinetic model parameters and the correlation coefficients are presented in Table 4. Figure 3.3 shows the pseudo-first order kinetics modeling of As(V) and As(III) adsorption on to IOCS and GFH. The equilibrium rate constants of pseudo-first order sorption (k_1) vary between 1.49×10^{-3} and 8.98×10^{-3} min^{-1}. The results show good correlation coefficients for both IOCS and GFH.

The equilibrium rate constant of the pseudo second-order kinetic are presented on Figure 3.4. The initial adsorption rate (*h*), the rate constants (k_2) and the correlation coefficients (R^2) were calculated and are presented in Table 3.4A and 3.4B.

Table 3.4: Effect of Ca^{2+} on adsorption kinetics parameters of As(III) (A) and As(V) (B) sorption by GFH and IOCS

A

		GFH + As(III)	GFH + Ca + As(III)	IOCS + As(III)	IOCS + Ca + As(III)
First order Kinetic	k_1 (min^{-1})	3.49 x10^{-3}	4.26 x10^{-3}	1.49 x10^{-3}	2.80 x10^{-3}
	qe	1.92	1.93	1.80	2.16
	R^2	0.94	0.98	0.96	0.99
Pseudo Second order kinetic	K_2 (gmg^{-1}min^{-1})	1.00 x10^{-5}	1.66 x10^{-3}	7.63 x10^{-4}	4.06 x10^{-4}
	qe	15.6	7.1	4.6	7.9
	h (mgg^{-1}min^{-1})	2.44 x10^{-2}	8.42 x10^{-2}	1.58 x10^{-2}	2.56 x10^{-2}
	R^2	0.99	0.98	0.91	0.98
Intraparticle diffusion	K_{id} (mgg^{-1}min$^{-1/2}$)	0.31	0.09	0.13	0.23
	Intercept	0.97	4.92	0.28	1.24
	R^2	0.93	0.93	0.96	0.93
Elovich	α (mgg^{-1}min^{-1})	0.08	12.22	0.15	0.04
	a (mgg^{-1}min^{-1})	0.38	1.25	0.89	0.51
	R^2	0.99	0.88	0.93	0.99

B

		GFH + As(V)	GFH + Ca + As(V)	IOCS + As(V)	IOCS + Ca + As(V)
First order Kinetic	k_1 (min^{-1})	8.89 x10^{-3}	6.59 x10^{-3}	2.73 x10^{-3}	2.79 x10^{-3}
	qe	1.87	1.76	2.16	1.75
	R^2	0.94	0.91	0.93	0.99
Pseudo Second order kinetic	K_2 (gmg^{-1}min^{-1})	2.26 x10^{-3}	4.44 x10^{-3}	1.09 x10^{-3}	1.12 x10^{-3}
	qe	7.9	8.0	6.0	5.9
	h (mgg^{-1}min^{-1})	1.42 x10^{-1}	2.85 x10^{-1}	3.97 x10^{-3}	3.98 x10^{-2}
	R^2	0.99	0.99	0.91	0.98
Intraparticle diffusion	K_{id} (mgg^{-1}min$^{-1/2}$)	0.18	0.13	0.15	0.13
	Intercept	4.38	5.53	2.09	2.45
	R^2	0.67	0.82	0.88	0.86
Elovich	α (mgg^{-1}min^{-1})	2.05	3.50	0.10	0.15
	a (mgg^{-1}min^{-1})	0.65	0.89	0.80	0.97
	R^2	0.82	0.93	0.97	0.96

Figure 3.5 shows the intraparticle diffusion plot for As(III) and As(V) adsorption by IOCS and GFH. The intraparticle diffusion rate constants k_{id} and the intercept were calculated from the slope of the plots. The intercepts of the plots are proportional to the thickness of the boundary layer (Acheampong et al., 2012) and varied between 0.28 and 5.53 mg.g^{-1}. The correlation coefficients (R^2) for adsorption of As(V) by IOCS and GFH were below 0.9.

Figure 3.6 shows the Elovich plot. The initial adsorption rate α and the desorption capacity β were obtained from the slope and the intercept, respectively. The correlation coefficients obtained indicate that the model fits the experimental data well, except for the adsorption of

As(V) by GFH without calcium addition and the adsorption of As(III) by GFH with addition of calcium.

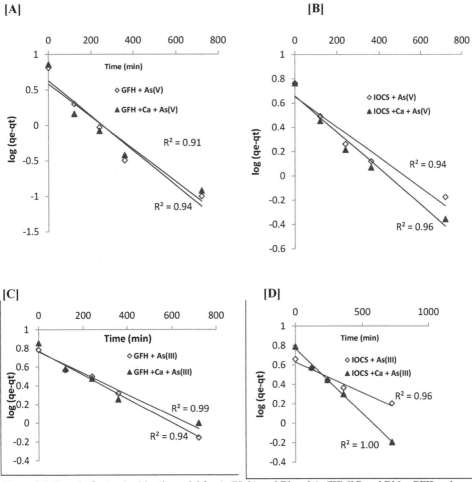

Figure 3.3: Pseudo-first order kinetic model for As(V) [A and B] and As(III) [[C and D] by GFH and IOCS. Model water composition: [As] = 3.88 mg/l, [Ca^{2+}] = 0 and 80 mg/l, [HCO$_3^-$] = 20 mg/l, pH 7

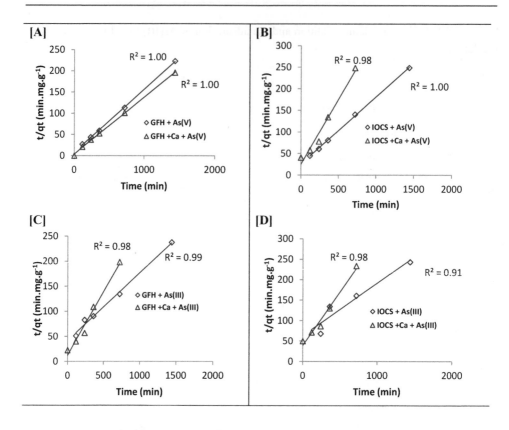

Figure 3.4: Pseudo-second order kinetic model for As(V) [A and B] and As(III) [C and B] by GFH and IOCS. Model water composition: [As] = 3.88 mg/l, [Ca^{2+}] = 0 and 80 mg/l, [HCO$_3^-$] = 20 mg/l, pH 7

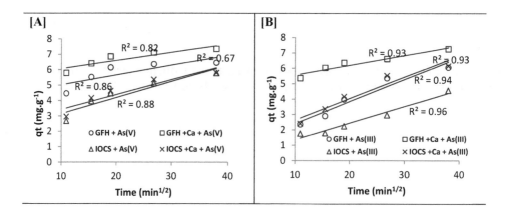

Figure 3.5: Intraparticle diffusion kinetic model for As(V) [A] and As(III)[B] by GFH and IOCS. Model water composition: [As] = 3.88 mg/l, [Ca^{2+}]= 0 and 80 mg/l, [HCO$_3^-$]= 20 mg/l, pH 7

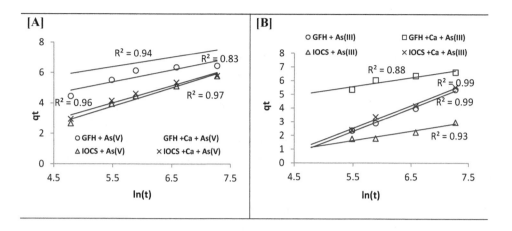

Figure 3.6: Elovich equation for As(V) [A] and As(III) [B] by GFH and IOCS. Model water composition: [As] = 3.88 mg/l, $[Ca^{2+}]$ = 0 and 80 mg/l, $[HCO_3^-]$ = 20 mg/l, pH 7

3.3.5 RSSCT experiments

Figure 3.7 shows the RSSCT breakthrough curves for As(III) and As(V) adsorption on IOCS and GFH at pH 7. After 10 hours of filter run and approximately 1040 Empty Bed Volumes (EBV) of calcium-free synthetic water filtered, the ratios of C/C_o for As(V) were 26% and 18% for filter columns with IOCS and GFH, respectively. The presence of 80 mg/l of calcium in model water dramatically changed the breakthrough pattern with a C/C_o of only 1% and 0.2% for IOCS and GFH, respectively. A similar behavior was observed for As(III); C/C_o of 69% and 60% for IOCS and GFH, respectively, in the absence of calcium and the ratio C/C_o of 19% and 0.8% in the presence of 80 mg/l of calcium, at pH 7, for IOCS and GFH, respectively.

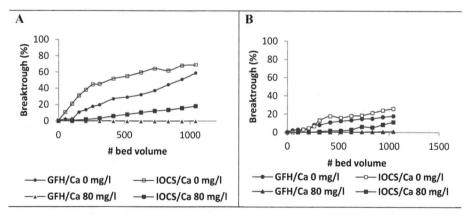

Figure 3.7: RSSCT breakthrough curve for As(III) (A) and As(V) (B) with IOCS and GFH. Model water: As(III) = 3.88 mg/l, pH 7, Ca = 0 and 80 mg/l, HCO_3^- = 20 mg/l, IOCS mass = 26.7 g, GFH mass = 33.6 g, Flow rate = 73.8 ml/min, EBCT = 0.35 min

3.4 Discussion

The effects of cations on the adsorption of As(III) and As(V) onto soil surfaces (oxisol, vertisol and alfisol) have been studied by Smith *et al.* (2002). They observed that the sorption of As(V) was enhanced by the presence of calcium in the solution more than sodium. They concluded that the effects of a cation may occur through specific sorption, leading to an increased positive charge on the surface of adsorbent. An increase in the valence of the cation makes the potential in the plane of sorption less negative, thereby increasing anion sorption. The effect of cation charge on surface potential was also studied by Bolan *et al.* (1993) for variably charged soils. These investigators found that specific sorption of calcium increased the surface positive charge and led to enhanced retention of SO_4^{2-}. The effect of calcium on As(V) adsorption by kaolinite has also been studied by Cornu *et al.* (2003) who found an increase in As(V) adsorption by increasing calcium concentration. It has also been reported that calcium can form complexes with arsenic in the forms of $CaH_2AsO_4^+$, $CaHAsO_4$, $CaAsO_4^-$ and $Ca_3(AsO_4)_2$ (Mironov et al. 1995) and the neutral forms of metal complexes appear to have higher affinity to the sorbent surfaces (Macnaughton and James 1974, Kinniburg and Jackson 1978). The effect of calcium on arsenic removal was also explained in the results obtained by Smith *et al.* (2002); they concluded that the presence of calcium increases the positive charges on the surface of the adsorbent and this increases the adsorption of negatives ions. This was confirmed by Bolan *et al.* (1993) who found that an increase in the valence of the cation increased anion sorption. Discussions on the results obtained in this study are as follow:

3.4.1 PHREEQC-2 modeling
The minerals predicted by PHREEQC-2 modelling all have negative saturation indices for both As(III) and As(V). These minerals are arsenolite (As_2O_3) and claudetite (As_4O_6) for As(III) with saturation indices -14.21 and -13.92, respectively. The predicted mineral for As(V) to precipitate in the solution was arsenic pentoxide (As_2O_5) but its saturation index was -25.67. This implies that no precipitation was expected in the model water for both As(III) and As(V).

3.4.2 Short term adsorption batch experiments
The results from short term batch experiments showed an enhancement in the removal of As(III) and As(V) by IOCS and GFH in the presence of calcium. The presence of $Ca_3(AsO_4)_2$ was predicted by the PHREEQC-2 modeling in the pH range studied, and it can be assumed that the adsorption of this species contributed to the better removal of arsenic by both IOCS and GFH. Ca^{2+} ions likely act as a bridging agent between the negatively charged IOCS/GFH surfaces and negatively charged As (V) ions at pH values above the pH_{pzc}. As(V) oxi-anions might have been adsorbed in the second adsorption sphere of the IOCS and GFH via a Ca^{2+}-bridge. At pH values below the pH_{pzc} (Table 3.1), the mechanism of As(V) removal by IOCS and GFH could be attributed to the combination of ligand exchange and electrostatic attraction, and consequently the beneficial effect of calcium was less pronounced.

Better performance of GFH in comparison to IOCS can be explained by the high purity, iron content, and porosity of GFH. Better removal of As(V) was also observed at high pH when calcium was present, contrary to the expectation that better adsorption would be observed at low pH due to the positively charged surface of adsorbent. The limited effect of pH on the As(III) removal efficiency by IOCS/GFH is in agreement with the results obtained by Kundu and Gupta (2006), who studied the removal of As(III) by iron oxide coated cement (IOCC) from aqueous solution. Results emerging from their study showed very limited effect of the pH of the solution (3.2-12.0) on the removal of As(III).

3.4.3 Batch isotherm experiments

The Freundlich isotherm constants (K) obtained (845-3357 (mg/g) (L/mg)$^{1/n}$) suggest that the adsorption capacity of the adsorbents (IOCS and GFH) were high. This is in agreement with results reported in the literature (Alduri 1996, Petrusevski et al. 2002). The highest adsorption capacities (Table 3.2) confirmed results from screening experiments where As(V) was better adsorbed than As(III), and more so if calcium was present in the model water (Figure 3.3). Relatively low values of the coefficient 1/n <1 (0.20-0.38) suggests that changes in the equilibrium concentration of arsenic will have a limited effect on the IOCS and GFH arsenic adsorption capacities.

3.4.4 Adsorption kinetic study

The pseudo-first order and pseudo-second order kinetics modeling of As(V) and As(III) adsorption onto IOCS and GFH were performed. In comparison to the pseudo-first order kinetics, the pseudo-second order kinetics fits the experimental data better in terms of correlations coefficients, except for the adsorption of As(III) and As(V) by IOCS without the addition of calcium. Thus, one can conclude that the adsorption of As(III) and As(V) onto GFH is likely a second order reaction with and without addition of calcium. The adsorption of As(III) and As(V) onto IOCS likely follows a first-order reaction, but when calcium is added, the reaction becomes pseudo-second order.

The intraparticle diffusion plot (Figure 3.5) showed the correlation coefficients (R^2) for adsorption of As(V) by IOCS and GFH below 0.9, which implies that the adsorption of As(V) does not follow the intraparticle diffusion model. The bigger the intercept, the greater the contribution of the surface sorption to the rate controlling step, as stated by Acheampong et al. (2012). However, the adsorption of As(III) fits the intraparticle diffusion model well with good correlation coefficients (> 0.9). Hence the main controlling process for As(III) adsorption is intraparticle diffusion while the surface adsorption contributes greatly to the adsorption of As(V).

3.4.5 RSSCT experiments

Results from RSSCT experiments support the results obtained in batch adsorption experiments, where the presence of calcium also had a beneficial effect on the adsorptive removal of As(III) and As(V) with IOCS and GFH. Results from the RSSCT experiments confirmed that GFH performed better than IOCS, and As(V) was better adsorbed compared to As(III).

3.5 Conclusions

- The presence of calcium in the model water enhanced As(III) adsorption on IOCS at all pH values studied (6, 7, 8), the effect of calcium on As(V) removal by IOCS was limited at pH values 6 and 7 while it strongly improved As(V) removal at pH 8. Similar results were obtained for As(III) and As(V) adsorption by GFH.
- The Freundlich isotherm constants (K) obtained (845-3357 $(mg/g)(L/mg)^{1/n}$) showed that adsorption capacity of IOCS and GFH was high, the highest K values were obtained when calcium was present in the model water.
- The adsorption of As(III) and As(V) onto GFH follows a second order reaction with and without addition of calcium while the adsorption of As(III) and As(V) onto IOCS follows a first-order reaction without calcium addition, and moves to the second reaction order kinetics when calcium is added.
- The main controlling process for As(III) adsorption is intraparticle diffusion while the surface adsorption contributes greatly to the adsorption of As(V).
- Results obtained in the RSSCT study confirmed the beneficial effect of calcium on As(III) and As(V) adsorption onto IOCS and GFH, which can significantly prolong the run length of adsorptive filters.

3.6 References

Acheampong M.A., Pereira J.P.C, Meulepas R.J.W and Lens. P.L.N, 2012, Kinetics modeling of Cu(II) biosorption on to cocunut shell and Moringa oleifera seeds from tropical regions, Environ. Techn. 33, 409-417

Alduri B. 1996 Adsorption modeling and mass transfer. In: G McKay (ed.) Use of Adsorbents for the Removal of Pollutants from Wastewaters, 138, CRC press, New York, p 136

Amy G., Chen H., Drizo A., Gunten U., Brandhuber P., Hund R., Chowdhury Z., Kommineni S., Sinha S., Jekel M., Banerjee K. 2005 Adsorbent Treatment Technologies for Arsenic Removal, AWWA research foundation and American water works association, Washington, D.C

Appelo C.A.J. and Postma D. 2005, Geochemistry, Groundwater and Pollution, 2nd edition, CRC Press, Taylor & Francis Group, 6000 Broken Sound Parkway NW, Suite 300

Benjamin M., Sletten R. and Bennet T. 1996 Sorption and Filtration of Metal using Iron-Oxide Coated Sand, Water Research Vol. 30 No. 11, 2609-2620

Bissen M., Gremm T., Köklü Ü, and Frimmel F.H. 2000 Use of the Anion-Exchange Resin Amberlite IRA-93 for the Separation of Arsenite and Arsenate in Aqueous Samples, Acta hydrochimica and hydrobiologica, volume 28, issue 1, 41-46

Bolan N.S., Syers J.K, Summer M.E. 1993 Calcium-induced sulfate adsorption by soils. Soil Sci. Soc. Am. J. 57, 691-696.

Chen H.W., Frey M.M., Clifford D., Mcneill L.S., and Edwards M. 1999 Arsenic Treatment Considerations. Journal of the American Water Works Association, 91, 74-85.

Chen, Z., Ma, W., Han, M., 2008. Biosorption of nickel and copper onto treated alga

(*Undaria pinnatifida*): Application of isotherm and kinetic models. *Journal of Hazardous Materials*, **155**(1-2):327-333.

Cornu S., Breeze D., Saada A. and Baranger P. 2003 The Influence of pH, Electrolyte, Type, and Surface Coating on Arsenic (V) Adsorption onto kaolinites, soil Sci.Soc.Am. J. 67,127-1132

Dzombak D.A. and Morel F.M.M. 1990 Surface Complexation Modeling: Hydrous Ferric Oxides. Wiley – Interscience Publication; John Wiley & Sons, Inc., New York

Faust. O.M. and Aly O.M. 1987 Adsorption Processes for Water Treatment Butterworths, Boston 134

Goldberg S., Johnston C.T. 2001 Mechanisms of Arsenic Adsorption on Amorphous oxides Evaluated Using Macroscopic Measurements, Vibrational Spectroscopy, and Surface Complexation Modeling, Journal of Colloid and Interface Science 234 204–216

Hameed, B.H., 2008. Equilibrium and kinetic studies of methyl violet sorption by agricultural waste. *Journal of Hazardous Materials*, **154**(1-3):204-212.

Hering J.G., Chen P.Y., Wilkie J.A., Elimelech M., and Liang S. 1996 Arsenic removal by ferric chloride. Jour. AWWA. 88, 155-167.

Ho Y.S., Wase D.A.J. and Forster C.F. 1995 Batch Nickel removal from aqueous solution by sphagnum moss peat, Water Research, 29 (5), 1327-1332

Ho Y.S. 2006, Review of second order models for adsorption systems. *J. Hazard. Mater.* 136, 681-689

Ho Y.S., and McKay G, 2000 The Kinetic of Sorption of Divalent Metal ions on to Spagnum Moss Flat, Water Resource, 34(3), 735

Ho Y.S. and Wang C.C. 2004, Pseudo-isotherms for the sorption of cadmium ion onto tree fern, *Process Biochem* 39, 761-765 .

Huang, W.W., Wang, S.B., Zhu, Z.H., Li, L., Yao, X.D., Rudolph, V., Haghseresht, F., 2008. Phosphate removal from wastewater using red mud. *Journal of Hazardous Materials*, 158(1):35-42.

Jeong Y., Fan M., Van Leeuwen J., Belczyk J.F. 2007 Effect of competing solutes on arsenic(V) adsorption using iron and aluminum oxides. Journal of Environmental Sciences 19 910-919

Kinnburg D.G., Jackson M.L. 1978 Adsorption of mercury (II) by iron hydrous oxide gel. Soil Science Society of American Journal 42, 45-47

Kundu S. and Gupta A.K. 2006 Adsorptive removal of As(III) from aqueous solution using iron oxide coated cement (IOCC): Evaluation of kinetic, equilibrium and thermodynamic models, Separation and Purification Technology 51, 65–172
Langmuir I., 1916 The constitution and fundamental properties of solids and liquids, J. Am. Chem. Soc. 38, 2221–2295.

Lagergren, S., 1898. About the theory of so-called adsorption of soluble substances. *Kungliga Svenska Vetenskapsakademiens. Handlingar*, **24**(4):1-39.

Liu D., Sansalone J.J. and Cartledge F.K. 2004 Adsorption characteristics of oxide coated buoyant media for storm water treatment I: batch Equilibria and Kinetic models, Journal of Environmental Engineering 130(4), 383-390

Macnaughton M.G., James R.O. 1974 Adsorption of aqueous mercury (II) complexes at the oxide/water interface. Journal of Colloid Interface Science, 47, 431-440

Masel R. 1996 Principles of Adsorption and Reaction on Solid Surfaces. Wiley, New York

Mcneill L.S. and Edwards M. 1997(a) Arsenic Removal during Precipitative Softening Journal of Environmental Engineering, 123, 453-460.

Mironov V.E., Kiselev V.P., Egyzarian M.B., Golovnev N.N., Pashkov G. 1995 Russian Journal of Inorganic Chemistry, 40, 1752-1733

Mohan D., Pittman J. 2007 Arsenic Removal from Water/Wastewater using Adsorbents-A critical review. Journal of Hazardous Materials 142(1-2) 1-53

Möller T. and Sylvester P. 2008 Effect of Silica and pH on Arsenic Uptake by Resin/Iron Oxide Hybrid media. Water Research 42(6-7),1760-1766.

Petrusevski B., Boere J., Shahidullah S.M., Sharma S.K., Schippers J.C. 2002 Adsorbent based point-of-use system for arsenic removal in rural areas, J Water SRT - Aqua. 51, 135-144

Petrusevski B., Sharma S.K., Schippers J.C., Shordt K. 2007 Arsenic in drinking water, Thematic overview paper 17, IRC International Water and Sanitation Centre

Rosa, S., Laranjeira, M.C.M., Riela, H.G., Fávere, V.T., 2008, Cross-linked quaternary chitosan as an adsorbent for the removal of the reactive dye from aqueous solutions. *Journal of Hazardous Materials*,

Scott K.N., Green J.F., Do H.D., and Mclean S.J. 1995 (a) Arsenic Removal by Coagulation. Journal of the American Water Works Association, 87, 114-126.

Scott M.J., and Morgan J.J. 1995 (b) Reactions at Oxide Surfaces, 1. Oxidation of As(III) by Synthetic Birnessite. Environmental Science and Technology, 29, 1898-1905.

Sperlich A., Werner A., Genz A., Amy G., Worch E. and Jekel M. 2005 Breakthrough behavior of granular ferric hydroxide (GFH) in fixed-bed adsorption filters: modeling and experimental approaches. Water Res. 39 (6), 1190–1198.

Sherman D.M. and Randall S.R. 2003 Surface complexation of arsenic(V) to iron(III) (hydr)oxides: Structural mechanism from ab initio molecular geometries and EXAFS spectroscopy, Geochimica et Cosmochimica Acta, Vol. 67, No. 22, 4223–4230

Smedley P. S. and Kinniburgh D. G. 2002 A review of the source, behavior, and distribution of arsenic in natural waters Appl. Geochem., 17, 517–568.

Smith, A.H., Lingas, E.O., Rahman, M. 2000 Contamination of drinking water by arsenic in Bangladesh: a public health emergency. Bulletin World Health Org 78(9) 1093–1103

Smith E., Naidu R. and Alston A.M. 2002 Chemistry of Inorganic Arsenic in soils: II. Effect of Phosphorus, Sodium, and Calcium on Arsenic Sorption, J.Environ. Qual. 31,557–563.

Smith S.D., Edwards M. 2005 The influence of silica and calcium on arsenate sorption to oxide surfaces. J. Water Supply Res. Technol. 54 201–211.

Tan, I.A.W., Ahmad, A.L., Hameed, B.H., 2008. Adsorption of basic dye on high-surface-area activated carbon prepared from coconut husk: Equilibrium, kinetic and thermodynamic studies. *Journal of Hazardous Materials*, **154**(1-3):337-346.

Tien C. 1994Adsorbate transport: Its Adsorption and Rate. In: Adsorption Calculations and Modeling. Butterworth-Heinemann series in chemical engineering. Series Editor: Howard Brenner. Publishers- Butterworth-Heinemann, Boston, USA, 81-82

Wakansi D., Horsfall J.M. and Ayabaemi I.S. 2006 Sorption Kinetics of Pb^{2+} and Cu^{2+} ions from aqueous solution by Nipah palm (Nypa fruticans Wurmb) shoot biomass. Electronic Journal of Biotechnology -ISSN: 0717-3458; vol.9 No5., 587-592.

Wan Ngah, W.S., Hanafiah, M.A.K.M., 2008. Adsorption of copper on rubber (*Hevea brasiliensis*) leaf powder: Kinetic, equilibrium and thermodynamic studies. *Biochemical Engineering Journal*, **39**(3):521-530.

Wang L., Sorg T., and Chen A. 2000 Arsenic Removal by Full Scale Ion Exchange and Activated Alumina Treatment Systems. AWWA Inorganic Contaminants Workshop, Albuquerque, New Mexico.

WHO 2004 Guidelines for Drinking water quality, Second edition Vol.2: Health Criteria and Other supporting information. World health organization, Geneva, Switzerland

Westerhoff P., Highfield D., Badruzzaman M. and Yoon Yeomin 2005 Rapid Small-Scale Column Tests for Arsenate Removal in Iron Oxide Packed Bed Columns, Journ. of Environmental Engineering, 262-271

Chapter 4: Effect of phosphate on Chromium and Cadmium removal from groundwater by iron oxide coated sand (IOCS) and granular ferric hydroxide (GFH)

Part of this chapter has been presented as:

V. Uwamariya, B. Petrusevski, P. Lens, G.Amy (2011), Effect of Phosphate on Chromium Removal from Groundwater by Iron Oxide based adsorbents, In proceeding of the *IWA Specialist Groundwater conference*, Belgrade 8-10 September 2011.

V. Uwamariya, B. Petrusevski, P. Lens, G. Amy (2012), Effect of water matrix on adsorptive removal of heavy metals from groundwater. In proceeding of the *IWA World Water Congress and Exhibition*, Busan (Korea) 16-21 September 2012

V. Uwamariya, B. Petrusevski, P. N.L. Lens and G. Amy (2013), Effect of Phosphate on Adsorptive Removal of Chromium and Cadmium from Groundwater by Iron Oxide Based Adsorbents, *Journal of Water Science and Technology*, submitted

Abstract

In this study, the effect of PO_4^{3-} on the adsorptive removal of Cr(VI) and Cd^{2+} was assessed using Iron Oxide Coated Sand (IOCS) and Granular Ferric Hydroxide (GFH) as adsorbents. Batch adsorption experiments and rapid small scale column tests (RSSCT) were conducted using Cr(VI) and Cd^{2+} containing model water at pH 6, 7 and 8.5. The best Cr(VI) and Cd^{2+} adsorption was observed at pH 6. GFH showed much better removal of Cr(VI) than IOCS, while IOCS removed Cd^{2+} better than GFH. Increasing PO_4^{3-} concentrations in the model water from 0 to 2 mg/L, at pH 6, induced a strong decrease in Cr(VI) removal efficiency from 93% to 24% with GFH, and from 24% to 17% with IOCS. A similar trend was observed at pH 7 and 8.5. An exception was for Cr(VI) removal with IOCS at pH 8.5, which was not affected by the PO_4^{3-} presence in model water.. The effect of PO_4^{3-} on Cd^{2+} is clearly seen at pH 6 because there is no precipitation of cadmium in the model water. . At pH 8.5, the precipitation is the main process as it represents around 70% of cadmium removal. . The isotherm constants *K* for different combinations confirm the inhibition of Cr(VI) and enhancement of Cd^{2+} adsorption with presence of PO_4^{3-}. The same conclusion is confirmed by the results from rapid small scale column tests. The mechanism of Cr(VI) adsorption by GFH and IOCS is likely a combination of electrostatic attraction and ligand exchange, while the mechanism of Cd^{2+} at lower pH of 6, was sufficiently energetic to overcome some electrostatic repulsion.

Keywords: Adsorption, Groundwater, Chromium, Cadmium, GFH, IOCS, Phosphate, pH

4.1 Introduction

Cadmium and chromium are among the most harmful metals to human and animals. As a result of inappropriate waste-disposal practices, chromium contamination of surface and groundwater has become a significant environmental problem (Palmer and Wittbrodt, 1991). Chromium can also occur naturally as a result of dissolved minerals from weathering of chromites and other chromium-bearing minerals present in bedrock and soil (Nriagu and Nieboer, 1988).

Chromium has two oxidation states in aqueous systems, namely Cr(VI) and Cr(III), that have different mobility and toxicity. Because of the low solubility of $Cr(OH)_3(s)$ and $(Cr,Fe)(OH)_3(s)$ precipitates, and its strong adsorption onto solids under slightly acidic to basic conditions, the mobility of Cr(III) in the aquatic environment is expected to be low. Cr(III) attains its minimum solubility in the pH range of natural waters (i.e. pH 7.5–8.5) (Sharma et al. 2008). Cr(VI) exists in solution as monomeric species/ions: $H_2CrO_4^{0}$, $HCrO_4^-$ (hydrogen chromate) and CrO_4^{2-} (chromate); or as the dimeric ion $Cr_2O_7^{2-}$ (dichromate only exists in very strongly acidic solution). In the normal pH range of groundwater (i.e. pH 6.5–8.5), divalent chromate CrO_4^{2-}, predominates. In dilute solutions (< 1 mg/L), CrO_4^{2-}, being negatively charged, does not complex with negatively charged surfaces (Pichel et al. 1976). Chromium toxicity depends on chemical speciation as reported by Sharma et al. (2008). Cr(VI) species are more soluble than Cr(III), and are much more toxic to microorganisms, plants, animals and humans. Cr(VI) produces liver and kidney damage, internal hemorrhage and respiratory disorders, dermatitis and skin ulceration. Sharma et al. (2008) also reported that chromium was classified as a human carcinogen (group A) by USEPA with a maximum contaminant level (MCL) for total chromium of 100 μg/L. The World Health Organization (WHO), EC water directives as well as Canada and Australia recommend the guideline value of 50 μg/L of total chromium (WHO, 2011).

Cadmium, as one of the toxic trace metals, can be introduced into groundwater through agricultural application of sewage sludge and fertilizers, and/or through land disposal of metal-contaminated municipal and industrial wastes (Hu, 1998). Cadmium has no essential biological function and is extremely toxic to humans. In chronic exposure, it also accumulates in the body, particularly in the kidneys and the liver. Acute poisoning from inhalation of fumes and ingestion of cadmium salts can also occur and at least one death has been reported from self-poisoning with cadmium chloride (Baldwin and Marshall, 1999). An investigation of human health risks shows that cadmium causes many diseases if inhaled in higher doses. The disease called Itai-Itai (Ouch-Ouch) is well documented in Japan. It was named Itai-Itai due to pain caused by the decalcification and final fracturing of bones symptomatic of cadmium poisoning (WHO, 1996). Based on the possible toxicity of cadmium, the WHO guideline value is 3μg/l as a safe concentration in drinking water (WHO, 2011).

Several methods for chromium and cadmium removal such as precipitation, electrochemical reduction, adsorption onto different media, ion exchange, solvent extraction, nanofiltration, reverse osmosis and biological removal have been applied (Sharma et al., 2008, Mayo et al., 2007, Hu et al., 2004 and Fendorf et al., 1997). However, these techniques have their own inherent limitations such as limited efficiency, sensitive operating conditions, production of secondary sludges that are costly to dispose, etc. (Maran and Protton, 1971). The adsorption process has many advantages such as: low cost of adsorbent, easy availability, possible utilization of industrial, biological and domestic waste as adsorbents, low operational cost, ease of operation compared to other processes, reuse of adsorbent after regeneration. The adsorption process also has high capacity for heavy metal removal over a wide range of pH and to a much lower level. It has the ability to remove complex forms of metals that are generally not possibly by other methods, it is environmentally friendly (Rao et al., 2010).The sorption of heavy metals by soils or metal oxides has been extensively studied (Christophi and Axe, 2000; O'Reilly et al., 2001). Results suggest that sorption appears to be a multi-step process involving an initial fast adsorption, followed by a slow adsorption and diffusion into solid particles. Adsorption of chromium and cadmium on different sorbents such as iron oxide, iron coated sand, iron coated activated carbon and granular ferric hydroxides (Driehaus et al., 1998) have also been investigated.

Raw water typically contains a mixture of many ions and compounds in addition to the target one. These ions may enhance adsorption, may act relatively independently or may interfere with one another. Thus, the aim of this study is to investigate the effects of PO_4^{3-} on adsorptive removal of Cr(VI) and Cd^{2+} by Iron Oxide Coated Sand (IOCS) and Granular Ferric Hydroxide (GFH) under different conditions.

4.2 Materials and Experimental Methods

4.2.1 Adsorbents

GFH was obtained from the manufacturer GEH Wasserchemie in Osnabruck (Germany). IOCS was obtained from the Dutch Water Company Vitens, from the water treatment plant Brucht that treats groundwater with high iron content. For the screening and adsorption isotherm experiments, pulverized IOCS and GFH (< 63µm) were used. For RSSCT, a 356-390 µm size fraction of IOCS and GFH was used. IOCS was ground with a mortal and a pestle which was followed by sieving to obtain the required size fraction. Grinding was done with care as to obtain only the coating part of IOCS (excluding the sand core itself) for both batch and RSSCT experiments. Table 4.1 summarizes the physical properties of GFH and IOCS coating.

Table 4.1: Physical characteristics of GFH and IOCS

	Grain size (mm)	BET Surface area (m^2/g)	pH_{ZPC}	Porosity (%)	Moisture content (%)	Bulk density (g/cm^3)
GFH	0.2-2.0	555	7.5-8.0	0.75-0.80	43-48	1.32
IOCS	1.4-5.6	261	6.9	0.53	13.3	1.05

4.2.2 Model water

Model (synthetic) water was prepared by mixing demineralised water with the required amount of Cr(VI) and Cd^{2+} prepared from stock solutions of 100 mg/L Cr(VI) in form of $Cr_2O_7^{2-}$, and cadmium standard solution 100mg/L from HACH. 100 mg/L HCO_3^- was added to increase model water buffering capacity. Subsequently, pH was adjusted to the required values using 1M HNO_3 or NaOH solutions. PO_4^{3-} was obtained from KH_2PO_4 from Merck.

4.2.3 Short batch experiments

For batch adsorption experiments, acid-cleaned and closed 500 ml plastic bottles, fitted with tubes for periodic sampling, were filled with model water and 0.1g/l of pulverized IOCS or GFH (grain size < 63µm) were added. Bottles were placed on an Innova 2100 rotary shaker at 100 rpm and kept at 20 ±1°C. All samples were filtered through a 0.45 µm membrane filter using a polypropylene syringe filter.

4.2.3 RSSCT experiments

In addition to batch adsorption experiments, RSSCT were conducted assuming constant diffusivity as shown by Equations (3.1) and (3.2) in Chapter 3, paragraph 3.2.3. The assumed parameters for pilot scale filter column were the same as the ones presented in Chapter 3.

4.2.4 Analytical methods

Total Cr and Cd were analyzed with an atomic absorption spectrometer (Thermo Elemental Solaar MQZe-GF 95) with an auto-sampler and a graphite furnace used as detector (AAS-GF). For dilution, acidified demineralised water was used. Nickel nitrate (50 g Ni/l) was used as matrix modifier. Samples were passed through a 0.45 μm filter and acidified with HNO_3 to pH below 2. Cr(VI) and Cd were analyzed according to the Standard Methods for Examination of Water and Waste Water 18th edition (1992). PO_4^{3-} was measured with the ascorbic acid spectrophotometer method at 880 nm.

4.2.5 Isotherm model

In this study the experimental results were fitted in to the Freundlich isotherm Equation 3, presented in Chapter 2, paragraph 3.2.5. The experiments were performed at pH values of 6 and 7 with initial Cr(VI) concentration of 3 mg/L and initial concentration of Cd^{2+} equals to 5 mg/l, The solution was buffered by adding 100 mg/l of $[HCO_3^-] = 100$ mg/L, The adsorbent dosage was 0.1- 2 g/l.

4.3 Results

4.3.1 Screening short batch experiments

Short batch experiments with contact time of 24 hrs were conducted to screen the stability of Cr(VI) and Cd^{2+} in the model water, as well as to examine the effect of variable concentration of PO_4^{3-} on their removal. Contrary to Cr(VI), Cd was not stable in the model water except at pH 6. For other pH values, Cd^{2+} precipitates out as is clearly shown in Table 4.2.

Figure 4.1[A] clearly shows that IOCS barely removes Cr(VI). Only around 20% of Cr(VI) was removed by IOCS, while GFH removed up to 93% of Cr(VI). Figure 4.1[A] also shows that the addition of just 0.5 mg/l of phosphate can dramatically reduce the removal of Cr(VI) with GFH for up to 45%. The increase in phosphate concentration in the model water to 1 and 2 mg/l further decreased efficiency of Cr(VI) removal with GFH at all pH values studied.

Table 4.2: Comparison of total removal, adsorption and precipitation of Cadmium

		Blank	Total removal				Adsorption			
			Cd^{2+} 0 mg PO_4^{3-}/l	Cd^{2+} 0.5 mg/l PO_4^{3-}	Cd^{2+} 1 mg/l PO_4^{3-}	Cd^{2+} 2 mg/l PO_4^{3-}	Cd^{2+} 0 mg/l PO_4^{3-}	Cd^{2+} 0.5 mg/l PO_4^{3-}	Cd^{2+} 1 mg/l PO_4^{3-}	Cd^{2+} 2 mg/l PO_4^{3-}
GFH	pH 6	0.0	67.2	69.5	86.7	88.2	67.2	69.5	86.7	88.2
	pH 7	8.5	79.7	88.5	91.8	90.8	71.2	80	83.3	82.3
	pH 8.5	69.5	92.1	99.8	99.7	97.4	22.6	30.3	30.2	27.9
IOCS	pH 6	0.0	94.4	97.6	99	100	94.4	97.6	99.0	100
	pH 7	8.5	97.7	98	99.1	100	89.2	89.5	90.6	91.5
	pH 8.5	69.5	99.3	97.7	98.8	100	29.8	28.2	29.3	30.5

Figure 4.1[B] shows the effect of PO_4^{3-} on cadmium removal. The percentage of cadmium adsorbed was calculated as the difference between the total removal and removal through

precipitation (blank without adsorbent). The cadmium species likely to precipitate out are $Cd(OH)_2$ and $CdCO_3$ with K_{ps} values of 7.2×10^{-15} and 1.0×10^{-12}, respectively.

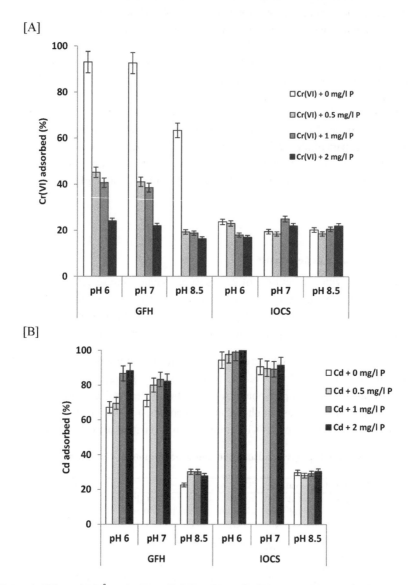

Figure 1: Effect of PO_4^{3-} and pH on Cr(VI) [A] and Cd [B] removal in batch adsorption experiments; Adsorbent: 0.1g/l GFH or IOCS ; CT=24 h; Model water: initial [Cr(VI)] = 0.5 mg/L; [Cd] = 0.3mg/l, [HCO$_3^-$] = 100 mg/L.

Figure 4.1[B] shows that Cd^{2+} was removed by both GFH and IOCS, but contrary to Cr(VI) removal, the removal was more effective when IOCS was used as adsorbent. The effect of PO_4^{3-} is clearly seen at pH 6 and 7 when there is no, or very limited precipitation in the solution. The presence of PO_4^{3-} increases the removal efficiency of Cd in contrast to the effect of phosphate on Cr(VI) removal. At pH 8.5, precipitation is the main Cd removal process, as it represents around 70% removal. The variation of PO_4^{3-} concentration in model water was also measured with time.

[A]

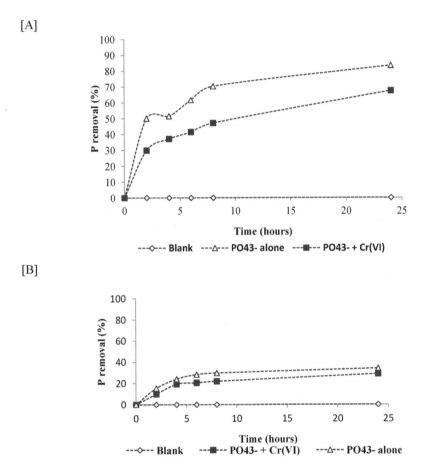

[B]

Figure 4.2: Effect of Cr(VI) on phosphate removal with GFH [A] and IOCS [B]; Model water: pH 7, $c(PO_4^{3-}) = 1mg/L$, $c(Cr(VI)) = 0$ or 0.5 mg/L; GHF / IOCS dosage: 0.1 mg/l

Figure 4.2 shows that PO_4^{3-} was partially removed by GFH and IOCS from model water that contained no Cr(VI). Presence of 0.5 mg/l of Cr(VI) in model water reduced phosphate removal efficiency by GFH and IOCS for approximately 16% and 5%, respectively. All the experiments were conducted with model water at pH 7. Similar to Cr(VI) removal, GFH was more effective than IOCS for PO_4^{3-} removal.

4.3.2 Adsorption isotherms

Adsorption isotherms were established for the adsorption of Cr(VI) by GFH at pH 7 and for the adsorption of Cd by IOCS at pH 6. Based on screening batch experiments Cr(VI) removal by GHF at pH 7, and Cd removal by IOCS at pH 6 were the most promising. .

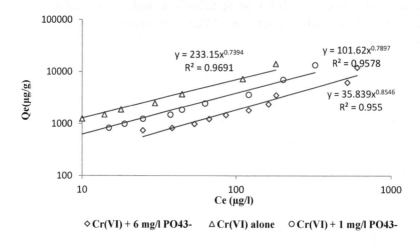

◇ Cr(VI) + 6 mg/l PO43-   △ Cr(VI) alone   ○ Cr(VI) + 1 mg/l PO43-

Figure 4.3: Freundlich adsorption isotherms for Cr(VI) adsorption on GHF at pH 7. Initial [Cr(VI)] = 3 mg/L; [PO$_4^{3-}$] = 0, 1 and 6 mg/L, [HCO$_3^-$] = 100 mg/L, GFH dosage = 0.1- 2 g/l.

Figure 4.3 shows Freundlich adsorption isotherms for Cr(VI) adsorption on GFH. The GFH adsorption capacity for Cr(VI) at Ce = 50 µgCr(VI)/l (corresponding to the WHO guideline for total chromium) was calculated to be 4.2 mg/g for model without PO$_4^{3-}$. The adsorption capacities decreased down to 2.2 mg/g and 1mg/g when 1 mg/L and 6 mg/L of PO$_4^{3-}$ were added to the model water.

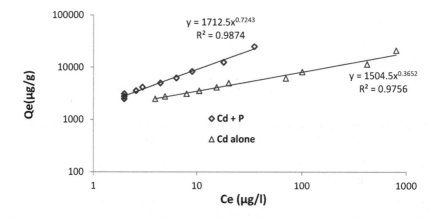

Figure 4.4: Freundlich adsorption isotherms for cadmium adsorption on IOCS and pH 6. Initial [Cd^{2+}] = 5 mg/l, [HCO$_3^-$] = 100 mg/L, [PO$_4^{3-}$] = 0, 1 and 6 mg/L, GFH dosage = 0.1- 2 g/l.

Figure 4.4 shows Freundlich isotherms for Cd^{2+} on IOCS. The IOCS adsorption capacity for an equilibrium concentration of 3µg/l of Cd (WHO, 2011) was calculated to be 2.2 for model water without PO_4^{3-}. The adsorption capacity calculated for the same equilibrium concentration of increased up to 3.8 mg/g when 1mg/l of PO_4^{3-} was added to the model water.

Table 4.3 shows the isotherm constants for Cr(VI) adsorption on GFH and Cd adsorption by IOCS from model water with different concentration of phosphate. The isotherm constant K_f for Cr(VI) removal from model water without phosphate is 2 and 6 times higher in comparison to model water with 1mg/L and 6mg/L of PO_4^{3-}, respectively.. The opposite is observed for the adsorption of Cd^{2+} by IOCS, the isotherm constant K_F being higher when 1 mg/l of PO_4^{3-} is added to the model water. The higher value of K_F suggests that the adsorption capacity of GFH to remove Cr(VI) is higher in the absence of PO_4^{3-}, while removal of Cd^{2+} by IOCS is better when PO_4^{3-} is added to the model water. This strongly supports an assumption that there is competition of PO_4^{3-} and Cr(VI) adsorption sites on the surface of the adsorbent, while the presence of PO_4^{3-} enhanced removal of Cd^{2+}. At an equilibrium concentration of 3 µg/l (WHO guideline for cadmium), the adsorption capacities of IOCS were 3.8 mg/g and 2.2 mg/g in the model water without, and with 1 mg/l of PO_4^{3-}, respectively.

Table 4.3: Freundlich isotherm constants for Cr(VI) adsorption onto GFH

Combination	K_F	n	R^2
3 mg/L Cr(VI) alone	233	1.35	0.97
3 mg/L Cr(VI) + 1mg/L PO_4^{3-}	102	1.27	0.96
3 mg/L Cr(VI) + 6mg/L PO_4^{3-}	36	1.17	0.95
5 mg/L Cd^{2+} alone	1505	2.74	0.98
5 mg/L Cd^{2+} + 1mg/L PO_4^{3-}	1713	1.38	0.99

4.3.3 RSSCT results

RSSCT results are presented in Figure 4.5. Figure 4.5 [A] shows that GFH adsorbs Cr(VI) much better than IOCS. After 10,000 empty bed volumes (EBV) of PO_4^{3-}-free model water filtered, 10% and 85% of Cr(VI) breakthrough (C/C_o) was observed in filter columns with GFH and IOCS, respectively. The presence of 1 mg/L of PO_4^{3-} in model water dramatically changed the breakthrough pattern with C/C_o of 74% for GFH after 10.000 EBV were filtered. When model water with Cr(VI) was filtered through column with IOCS, the 57% and 73% breakthrough was observed for model water without and with 1 mg/l of phosphate after treatment of about 4,000 EBV. Thus, results from RSSCT support the results from batch adsorption experiments that show much better removal of Cr(VI) with GFH. RSSCT also confirmed that GFH and IOCS adsorption capacity to remove Cr(VI) is reduced in the presence of PO_4^{3-} (Figure 5 [A]).

[A]

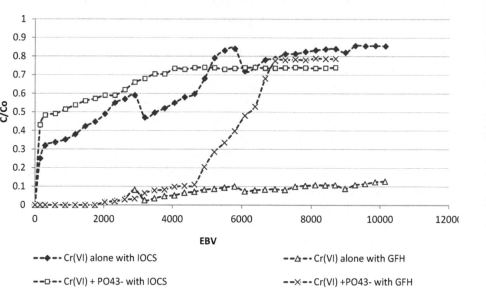

[B]

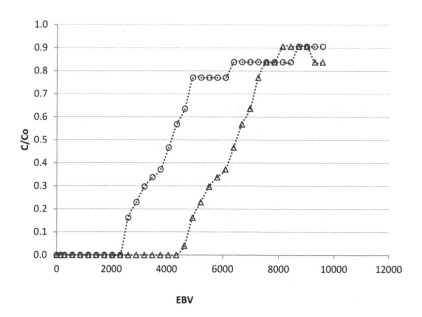

[C]

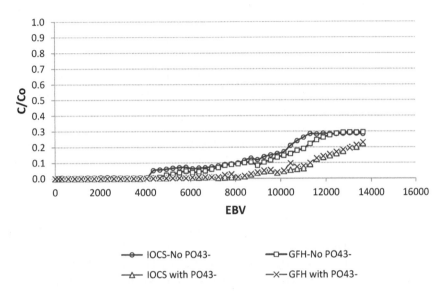

Figure 4.5: RSSCT breakthrough curves for Cr(VI) [A] , PO_4^{3-} [B]) and Cd^{2+} removal [C] with GFH and IOCS. Model water, pH 7, $c(HCO_3^-)$ = 100 mg/L, $c(Cr(VI))$ = 0.1 mg/L, $c(PO_3^{4-})$ = 0 and 1mg/L, Filtration rate = 46.6m/h, EBTC = 0.35 min.

Figure 4.5[B] shows that the capacity of GFH to remove PO_4^{3-} was higher than the capacity of IOCS. 80% of PO_4^{3-} breakthrough was obtained after treatment of about 4000 EBV for IOCS while the phosphate breakthrough of 90% was obtained after treating approximately 8000 EBV for GFH in the model water with 0.1mg/L of Cr(VI).

RSSCT results for cadmium removal with IOCS and GFH are presented on Figure 4.5[C]. After treatment of about 14,000 empty bed volumes (EBV) of PO_4^{3-}-free model water, the adsorbents were partially exhausted and comparable breakthrough of 30% was observed in filter columns with GFH and IOCS. The presence of 1 mg/L of PO_4^{3-} in the model water changed the breakthrough pattern with C/C_0 of 25% for GFH and IOCS after 14,000 EBV were filtered.

4.4 Discussion

In comparison with the results obtained in this study, previous researches have also shown competitive adsorption of chromium with different anions by showing that the presence of orthophosphate prevented the adsorption of Cr(VI) most likely by competition for adsorption sites. Consistent with these findings, Bartlett and Kimble (1976) reported the capability of KH_2PO4 to extract Cr(VI) soil. Tzou et al. (2003) also showed that phosphate (P) and organic ligands could influence Cr(VI) retention by the soil components. Phosphate (P) not only competed with Cr(VI) for surface sites, but also resulted in releasing sorbed Cr(VI).

The obtained results in this study are supported by previous results which have also shown that high concentrations of PO_4^{3-} in synthetic water decrease adsorption capacity of iron oxide (Gao and Mucci, 2001, Zhang et al. 2004, Manning and Golberg, 1996, Chowdhury and Yanful, 2010). Zhang et al. (2004) showed that addition of 6.5 mg/l PO_4^{3-} can lower arsenate adsorption on iron ore (5 g/l) by 30-50% at pH 7, and Gao and Mucci (2001) reported that the surface complexation of arsenate on goethite decreased in proportionality to the amount of phosphate present in solution. Chowdhury and Yanful (2010) showed that, at a fixed adsorbent (mixed magnetite-maghemite nanoparticles) concentration of 0.4 g/l, arsenic and chromium uptake decreased with increasing phosphate concentration. Nano-size magnetite-maghemite mixed particles adsorbed less than 50% arsenic from synthetic water containing more than 3 mg/l phosphate and 1.2 mg/l of initial arsenic concentration, and less than 50% chromium from synthetic water containing more than 5 mg/l phosphate and 1.0 mg/l of chromium (VI).

The previous findings also support the results obtained in this study on the effect of PO_4^{3-} on Cd^{2+} removal. Wang and Xing (2002) reported that cadmium adsorption by goethite was almost totally adsorbed at high pH, with the initial cadmium concentration of 10^{-5} M. In comparison with the results obtained in this study, one can assume that not all the cadmium was adsorbed but some amount precipitated out. Wang and Xing (2002) also found that presence of shifted the pH value at which Cd was adsorbed. After 15 min reaction, nearly all Cd were adsorbed at pH >7.2 in the absence of PO_4^{3-}, and at pH > 5.5 in the presence of PO_4^{3-} (Wang and Xing, 2002). Xiong (1995) also found that orthophosphate increased Cd^{2+} adsorption and decreased its desorption from three representative soils. Significant Cd^{2+} sorption by the phosphate-free oxide sample occurred at pH levels below the oxide's zero point of charge (pH_{ZPC}), indicating that the sorption was sufficiently energetic to overcome some electrostatic repulsion. This may explain why Cd^{2+} was well adsorbed at pH 6, which is below the point of zero charge of IOCS and GFH. PO_4^{3-} enrichment of the oxide enhanced its capability to retain Cd^{2+}, and lowered the average pH of the sorption edge (Kuo and McNeal, 1884). Several studies revealed that PO_4^{3-} pretreated goethite enhanced metal ions adsorption. Cd^{2+} adsorption increased with increasing pH, temperature and phosphate concentration. It was found that PO_4^{3-} formed both outer and inner sphere complexes via metal and ligand-like adsorption (Zaman *et al.* (2009, Venema *et al.*, 1997; Collins *et al.*, 1999; Wang and Xing, 2002). With the increase of PO_4^{3-}, more effective Cd^{2+} adsorption has been reported (Jie *et al.*, 2000), but depending on the way goethite was pretreated with oxalic acid, affinity for Cd^{2+} Cd^{2+} varied (Zhang *et al.*, 2001). Bolan et al. (1996) studied the effect of inorganic ions on the adsorption of cadmium by soil and they find that chloride decreases the adsorption of Cd^{2+} which is attributed to the complex formation of Cd^{2+} as $CdCl^+$ and $CdCl^+$, which are less readily adsorbed than Cd^{2+}. Sulphate and nitrate caused no significant change in Cd^{2+} adsorption whereas HPO_4^{2-} increased the adsorption of Cd^{2+} (Chun-yan, et al., 2009). The cadmium ion sorption decreased with an increase of phosphate sorption when adsorbed PO_4^{3-} was less than 40 mg·kg^{-1} in red soil and 37 mg·kg^{-1} in cinnamon soil; the average of adsorption ratio declined by 16% and 10%, respectively, as compared to no PO_4^{3-} application. Moreover, cadmium ion sorption increased with increasing PO_4^{3-} sorption when adsorbed PO_4^{3-} was higher than 79 mg·kg-1 in red soil and 70 mg·kg-1 in cinnamon soil. The adsorption ratio increased by 11% and 6% on an average, respectively as compared with the control.

In regards with the mechanism, Fendorf et al. (1997) showed that three different surface complexes exist on goethite: a monodentate complex, a bidentate-binuclear complex, and a bidentate-mononuclear complex. At low surface coverage, the monodentate complex was favored while at higher coverages the bidentate complexes were more prevalent. The bidentate-binuclear complex appears to be in the greatest proportion at these highest surface coverages. Tejedor-Tejedor and Anderson (1990) reported that PO_4^{3-} produces three different types of complexes on iron oxide (i.e., goethite, hematite, etc) surfaces: protonated $((FeO)_2(OH)PO)$, nonprotonated bridging bidentate $((FeO)_2PO_2)$ and nonprotonated monodentate $((FeO)PO_3)$. Monoprotonated binuclear phosphate complex on magnetite surface at pH 3 also have been reported by Daou et al. (2007). Thus similar dominant species of Cr(VI) and PO_4^{3-} and their affinity to iron oxide surface may be the reason for less effective removal of Cr(VI) in the presence of phosphate.

The mechanism of Cr(VI) adsorption by GFH and IOCS is likely a combination of electrostatic attraction and ligand exchange, as suggested by Hu et al. (2004) who studied the adsorption of Cr(VI) on magnetite. Chowdhury et al. (2010) and Hu et al. (2005) also suggested that electrostatic attraction is the key mechanism of chromium removal from aqueous solutions by maghemite $(g-Fe_2O_3)$, and that the process is highly dependent on initial chromium concentration, pH and temperature. Khaodhiar et al. (2000) also studied Cr(VI) adsorption and equilibrium modelling on IOCS. They found that chromate was weakly adsorbed or formed an outer-sphere surface complex ofthe IOCS surface. This can probably suggest why Cr(VI) was less effectively adsorbed by IOCS compared to GFH. Cr(VI) was probably adsorbed on GFH through inner-sphere surface complexes (strong complexes). This conclusion is supported by the results obtained by Hsia et al. (1993), who investigated the adsorption of Cr(VI) on hydrous iron oxide. According to their results, an inner-sphere coordination of chromate to the iron oxide surface was identified from EDAX and FTIR results. The isoelectric point of the system containing more chromate was a few pH units lower than that of the system containing less, or no chromate. This phenomenon was attributed to an increase in the negative charge of the iron oxide surface as a result of adsorption of chromate.

4.5 Conclusions

- In this study, IOCS and GFH showed the potential to adsorb Cr(VI), Cd^{2+} and PO_4^{3-} through the combination of electrostatic and complexation mechanisms. GFH showed higher adsorption capacity for both Cr(VI) and PO_4^{3-} than IOCS, but IOCS showed better removal of Cd^{2+}.

- For both adsorbents; better removal of Cr(VI), Cd^{2+} and PO_4^{3-} was observed at low pH value (pH 6). For Cd^{2+}, the precipitation of $Cd(OH)_2$ and $CdCO_3$ occurred at higher pH values of 7 and 8.5, even predominating over adsorption at pH 8.5. Better adsorption of Cr(VI) and PO_4^{3-} at low pH value could be explained by electrostatic attraction forces due to the positively charged adsorbent surface.

- Presence of PO_4^{3-} decreased the adsorption capacities of Cr(VI) and vice versa. Cr(VI) and PO_4^{3-}, being both negatively charged and having both affinity for iron oxides, compete for the surface adsorption sites on GFH and IOCS. Presence of PO_4^{3-} increased the adsorption capacities of Cd^{2+} by increasing the number of negative charges at the surface of the adsorbents. Thus, Cr(VI) removal by GFH and IOCS from contaminated groundwater is less favorable when PO_4^{3-} is present, while the opposite is observed when PO_4^{3-} is added to groundwater contaminated by Cd^{2+}.

4.6 References

Baldwin D.R, Marshall WJ. 1999 Heavy metal poisoning and its laboratory investigation (review article). Annals of Clinical Biochemistry; 36: 267-300. Available from URL: http://www.leeds.ac.uk/acb/annals/annals_pdf/May99/267.pdf

Bartlett, R.J., Kimble, J.M., 1976. Behavior of chromium in soils: II. Hexavalent forms. J. Environ. Qual. 5, 383e386

Bolan N. S., Khan M. A. R. and Tillman R. W., 1996, Effects of Inorganic Anions on the Adsorption of Cadmium By Soils, TEG Conference Proceedings, p 97

Chowdhury S.H. and Yanful E.K. 2010, Arsenic and chromium removal by mixed magnetite-maghemite nanoparticles and the effect of phosphate on removal, *Journal of Environmental Management* 91, 2238-2247

Christophi C.A., Axe L., 1999. Competition of Cd, Cu, and Pb adsorption on goethite. *Environ. Engg.*, Vol. 126, No.1, pp. 66-74.

Chun-yan G., Ying W., Ming-gang X., Shi-wei Z., Fen-tao L.V., Miao-miao C. 2009, Phosphate Adsorption and Its Effect on Adsorption-Desorption of Cadmium in Red Soil and Cinnamon Soil, Journal of Agro-Environment Science;2008-06

Collins C.R., Ragnarsdottir K.V., Sherman D.M., 1999, Effect of inorganic and organic ligands on the mechanism of cadmium sorption to goethite. *Geochimica et Cosmochimica Acta*, Vol. 63, pp. 2989–3002.

Daou, T.J., Begin, C.S., Greneche, J.M., Thomas, F., Derory, A., Bernhardt, P., Legare, P., Pourroy, G., 2007. Phosphate adsorption properties of magnetite-based nanoparticles. *Chem. Mater.* 19, 4494e4505

Driehaus, W., Jekel, M., Hildebrandt, U., 1998. Granular ferric hydroxide – a new adsorbent for the removal of arsenic from natural water. *J.Water SRT Aqua.* 47, 30–35.

Fendorf, S., Eick, M.J., Grossl, P., 1997. Arsenate and chromate retention mechanisms on goethite. 1. Surface structure. *Environ. Sci. Technol.* 31, 315-320.

Gao, Y., Mucci, A., 2001. Acid base reactions, phosphate and arsenate complexation, and t   heir competitive adsorption at the surface of goethite in 0.7 M NaCl solution. *Geochim. Cosmochim. Acta* 65 (14), 2361-2378.

Heijman S.G.J., Paassen A.M., Meer W.G., Hopman J.R., 1999. Adsorptive removal of natural organic matter during drinking water treatment. *Water Sci. Technol.*, Vol. 40, pp. 183- 190.

Hsia, T. H., Lo S. L., Lin C. F. and Lee D. Y. 1993, Chemical and Spectroscopic Evidence for Specific Adsorption of Chromate on Hydrous Iron Oxide, *Chemosphere*, Vol.26, No.10, pp 1897-1904

Hu H. 1998, Chapter 397: Heavy metal poisoning. In: Fauci AS, Braunwald E, Isselbacher KJ, Wilson JD, Martin JB, Kasper DL, Hauser SL, Longo DL (eds). Harrison's principles of Internal medicine. 14th ed. New York: McGraw-Hill; , pp2564-2569.

Hu, J., Lo, I.M.C., Chen, G., 2004. Removal of Cr(VI) by magnetite nanoparticle. *Water Sci. Technol.* 50 (12), 139e146.

Hu, J., Guohua Chen, G., Lo, I.M.C., 2005. Removal and recovery of Cr(VI) from wastewater by maghemite nanoparticles. *Water Res.* 39, 4528e4536.

Jie X.L., Liu F., Li X.Y., 2000. Effect of phosphate sorption on surface electrochemical properties of goethite and secondary sorption of zinc (In Chinese). *J. Henan Agricultural University*, Vol. 34, pp.109–121.

Khaodhiar, S., Azizian, M.F., Osathaphan, K., Nelson, P.O., 2000. Copper, chromium, and arsenic adsorption and equilibrium modeling in an iron-oxide-coated sand, background electrolyte system. *Water Air Soil Poll* 119, 105–120.

Kuo S. and McNeal B.L., Effects of pH and Phosphate on Cadmium Sorption by a Hydrous Ferric Oxide, *Soil Science Society of America Journal*, Vol. 48, *pp. 1040-1044*

Manning, B.A., Goldberg, S., 1996. Modeling competitive adsorption of arsenate with phosphate and molybdate on oxide minerals. Soil Sci. Soc. Am. J. 60, 121-131.

Maran S.H., Protton C.F., 1971. Principles of physical chemistry, 4th edition, The Macmillan Company, New York, Collier-Macmillan Ltd., London.

Mayo, J.T., et al., 2007. The effect of nano-crystalline magnetite size on arsenic removal. *Sci. Technol. Adv. Mater.* 8, 71-75.

Masel R. (1996) Principles of Adsorption and Reaction on Solid Surfaces. Wiley, New York

Nriagu, J.O., Nieboer, E., 1988. Historical Perspectives. IN: Chromium in the Natural and Human Environments, vol. 20. John Wiley & Sons, New York, pp. 1-20.

O'reilly S.E., Hochella Jr. M.F., 2003. Lead sorption efficiencies of natural and synthetic Mn and Fe-oxides. *Geochim.Cosmochim. Acta.* Vol. 67, pp. 4471–4487.

Palmer, C.D., Wittbrodt, P.R., 1991. Processes affecting the remediation of chromium-contaminated sites. *Environ. Health Perspect.* 92, 25-40.

Pichel, L., Butler, G. C. & Hoffman, I. (1976) Effect of Chromium in the Canadian Environment. National Research Council, Ottawa.

Rao K.S., Anand S., Venkateswarlu P., 2010. *Psidium guvajava l* leaf powder – A potential low cost biosorbent for the removal of cadmium(II) from wastewater. *Adsorption Science and Technology,* Vol. 28, No. 2, pp. 163-178.

Sen T.K., Mahajan S.P., Khilar K.C., 2002. Sorption of Cu2+ and Ni2+ on iron oxide and kaolin and its importance on Ni^{2+} transport in porous media. *Colloids Surfaces A: Physicochem.Eng. Aspects*, Vol. 211, pp.91–102.

Sharma S.K., Petrusevski B., and Amy, G. 2008 Chromium removal from water, Journal of *Water Supply: Research and Technology - AQUA* 57 (8), 541-553.

Tejedor-Tejedor, M.I., Anderson, M.A.,1990. Protonation of phosphate on the surface of goethite as studied by CTR-FTIR and electrophoretic mobility. *Langmuir* 6, 602-611.

Tzou, Y.M., Wang, M.K., Loeppert, R.H., 2003. Effects of phosphate, HEDTA, and light sources on Cr(VI) retention by goethite. Soil Sediment Contam. Int. J. 12 (1), 69-84.

Venema P., Hiemstra T., Weidler P.G., Riemsdijk W.H.V., 1998. Intrinsic proton affinity of reactive surface groups of metal (hydr)oxides: Application to iron (hydr)oxides. *J.*

Colloid Inter Sci, Vol.198, pp. 282-295.

Wang K, Xing B. 2002 Adsorption and desorption of cadmium by goethite pretreated with phosphate, *Chemosphere* 48, 665–670

Westerhoff P., Highfield D., Badruzzaman M. and Yoon Yeomin 2005 Rapid Small- Scale Column Tests for Arsenate Removal in Iron Oxide Packed Bed Columns, *Journ. of Environmental Engineering,* 262-271

WHO, 1996, *Guidelines for drinking water quality*, Second edition, volume 2, Health criteria and other supporting information

WHO 2004 Guidelines for Drinking Water Quality: First Addendum to Third Edition. Vol. 1 Recommendations, World Health Organization, Geneva, Switzerland.

Xiong L.M. 1995, Influence of phosphate on cadmium adsorption by soils, *Fertilizer research 40, pp 31-40*

Zaman M.I., Mustafa S., Khan S., Xing B., 2009. Effect of phosphate complexation on Cd^{2+} sorption by manganese dioxide (β-MnO_2). *J. Colloid Interf. Sci.,* Vol. 330, No.1, pp. 9-19.

Zhang G.Y., Dong Y.Y., Li X.Y., 2001. Effects and mechanisms of oxalate on Cd(II) sorption on goethite at different pH and electrolyte concentration (In Chinese). *Plant Nutrition and Fertilizer Science*, Vol. 7, pp. 305–310.

Zhang W., Singh, P., Paling, E., Delides S., 2004. Arsenic removal from contaminated water by natural iron ores. Minerals Eng. 17, 517-524.

Chapter 5: Effect of Calcium on adsorptive removal of Copper and Cadmium by Iron oxide coated sand and granular ferric hydroxide

Part of this chapter has been presented as:

V. Uwamariya, B. Petrusevski, P. Lens, N.S. Slokar, N. Stanič, G.Amy (2011), Removal of Heavy Metals (from groundwater) by Iron Oxide Coated Sand, In proceeding of the 2nd IWA Development Congress and Exhibition in Kuala Lumpur (Malaysia), November 21-24, 2011.

V. Uwamariya, B. Petrusevski, P. Lens, G. Amy (2012), Effect of water matrix on adsorptive removal of heavy metals from groundwater. In proceeding of the *IWA World Water Congress and Exhibition*, Busan (Korea) 16-21 September 2012

V. Uwamariya, B. Petrusevski, P. N.L. Lens and G. Amy (2013), Effect of Calcium on Adsorptive Removal of Copper and Cadmium from Groundwater by Iron Oxide Based Adsorbents, *Journal of Water Supply: Research and Technology-AQUA*, submitted

Abstract

In this study, iron oxide coated sand and granular ferric hydroxide were used to study the effects of pH and Ca^{2+} on the adsorptive removal of Cu^{2+} and Cd^{2+}. Batch adsorption experiments and kinetics modelling were performed. It was observed that Cu^{2+} and Cd^{2+} were not stable at the pH values considered, and the precipitation was predominant at higher pH values, especially for Cu^{2+}. The increase in Ca^{2+} concentration also increased the precipitation of Cu^{2+} and Cd^{2+}. It was also observed that Ca^{2+} competes with Cu^{2+} and Cd^{2+} for surface sites of the adsorbent. The presence of calcium diminishes the number of adsorption sites of iron oxide coated sand (IOCS) and granular ferric hydroxide (GFH) resulting in lower removal of cadmium and copper. Freundlich isotherms of cadmium removal by IOCS showed that the adsorption capacity of IOCS to remove Cd decreased when calcium was added to the model water. The kinetics modelling revealed that the adsorption of Cd onto IOCS is likely a second-order reaction.

Key words: Adsorption, Copper, Cadmium, Calcium, GFH, IOCS

5.1 Introduction

Access to safe drinking water is essential for life and may be regarded as a basic human right. However, water may be contaminated by chemical contaminants such as heavy metals. Heavy metal ions such as Cu^{2+}, Zn^{2+}, Fe^{2+}, are essential micronutrients for plant metabolism but when present in excess become extremely toxic (Williams et al., 2000).

Cadmium and copper are toxic for humans and animals at high concentration. Cadmium and copper metals can appear in leachates from urban and industrial waste. The migration of the leachate containing these metals through soil could lead to groundwater pollution. Being one of the most toxic heavy metals (Izanloo and Nasseri, 2005), cadmium is attracting wide attention of researchers. In fact, cadmium has been reported to be highly toxic because its homeostatic control in human body does not exist. The harmful effects of cadmium include a number of acute and chronic disorders, such as the "itai-itai" disease, renal damage, emphysema, hypertension, and testicular atrophy (Leyva-Ramos et al., 1997). Based on the possible toxicity of cadmium, the WHO guideline value is 3 µg/l as safe concentration in drinking water (WHO, 2004). Copper is essential to human health and can be found in many kinds of food, in drinking water and in air. However, too much copper can still cause eminent health problems. Long-term exposure to copper can cause irritation of the nose, mouth and eyes and it causes headaches, stomachaches, dizziness, vomiting and diarrhea. High uptakes of copper may cause liver and kidney damage and even death. Therefore, the concentration of copper must be reduced to levels that satisfy the environmental regulations for various bodies of water (Huang et al., 2006). The maximum concentration of copper in drinking water is 2.0 mg/l (WHO, 2004).

Several technologies have been used to remove heavy metals from groundwater, including coagulation, precipitation, adsorption, ion exchange, membrane filtration, as well as in situ and biological processes (Pal 2001, Petrusevski et al. 2002, Faust and Aly, 1998). However, chemical precipitation and most of other methods become inefficient when heavy metals are present in trace concentrations. Adsorption is one of few alternative treatment technologies available for such situations.

Recent studies have shown that quartz sand and other filter media coated with iron, aluminium, or manganese (hydro)oxides, or oxy-hydroxide are effective and inexpensive adsorbents for several metals (Khaodhiar et al. 2000, Petrusevski et al. 2002, Amy et al. 2005). However, different factors including the nature and characteristics of the adsorbate, ionic concentration, presence of organic matter, pH and temperature affect the effectiveness of the metal adsorption (Sharma et al. 2002, Abdu Salam and Adekola 2005). Calcium is normally an abundant element in groundwater and its competition with other metals in the adsorption process has been demonstrated previously (Kiewiet and Wei-Chun 1991,Escrig and Morell 1998, Temminghoff et al. 2005, Mustafa et al. 2004,   Hashim and Chu 2004, Benaissa and Benguella 2004).

The goal of this study was to screen the effect calcium on the adsorptive removal of cadmium and copper from groundwater by iron oxide coated sand (IOCS) and granular ferric hydroxide (GFH). Understanding the effects of the water quality matrix on the ability of iron (hydr)oxide adsorbents to remove heavy metals from contaminated waters will significantly improve the ability to optimize the treatment process.

5.2 Methodology

5.2.1 Experimental
GFH was obtained from the manufacturer GEH Wasserchemie in Osnabruck (Germany). IOCS was obtained from the Dutch Water Company (Brucht water treatment plant) that treats groundwater with high iron content. For batch experiments, a representative sample of pulverized IOCS and GFH (< 63µm) was used. The physical characteristics of the adsorbents IOCS and GFH are shown in Table 3.2 of the Chapter 3.

For batch adsorption experiments acid-cleaned and closed 500 ml plastic bottles, fitted with tubes for periodic sampling, were filled with the model water and the selected dosage of pulverized IOCS or GFH was added. Subsequently, the pH was adjusted to the required values using 1M HNO_3 or NaOH solutions. Bottles were placed on an Innova 2100 rotary shaker at 100 rpm and kept at 20 ($\pm$ 1)°C. All samples were filtered through a 0.45 µm membrane filter using a polypropylene syringe.

5.2.2 Analytical methods
Different metals were analyzed with an atomic absorption spectrometer (Thermo Elemental Solaar MQZe-GF 95) with an auto-sampler and a graphite furnace used as detector (AAS-GF). Calcium was analyzed with an atomic absorption spectrophotometer-Flame (Perkin-Elmer model AAnalyst 200).

5.2.3 Models applied
The PHREEQC-2 hydro geochemical model (Appelo and Postma 2005), the isotherm model and the adsorption kinetics models (Pseudo-first order kinetic model, Pseudo-second order

kinetic model, Intraparticle diffusion model and The Elovich equation) are all described in Chapter 3, paragraph 3.2.5 (Chen *et al.*, 2008; Hameed, 2008; Huang *et al.*, 2008; Wan Ngah and Hanafiah, 2008; Rosa *et al.*, 2008; Tan *et al.*, 2008, Lagergren, 1989, Ho, 2004, Ho, 2006, Acheampong et *al.*, 2012 and Ho and McKay, 2000).

5.3 Results

5.3.1 Stability tests

Prior to assess the effect of calcium on the adsorptive removal of Cu and Cd by IOCS and GFH, their stability in the model water was assessed. The objective of the experiments was to find out the effect of the contact time, pH and calcium concentrations on copper and cadmium stability in the solution. Model water with different pH values (6, 7 and 8) and calcium concentrations (0, 20, 40 and 60 mg/l) was used. Experiments were run for 24 hours.

Figure 5.1 shows that there is a decrease in dissolved Cu concentration in the solution due its precipitation as $Cu_2(OH)_2CO_3$ and $Cu_3(OH)_2(CO_3)_2$ compounds (calculated by Phreeqc interactive) at all pH values studied. Besides, an increase in calcium concentration resulted in a decrease in the Cu concentration in the solution. The decrease in dissolved Cu concentration increased as the pH was increased, the perecentage of precipitated Cu was low at pH 6 and without the addition of calcium (1.25%) and it was high at pH 8 with addition of 60 mg/l of Ca (84%). Figure 5.1 also shows that there is no significant decrease of the dissolved Cd concentration in the model water at pH = 6 for all calcium concentrations considered (20, 40 and 60 mg/l). Like for Cu, the increase in calcium concentration increased the precipitation of cadmium, the highest precipitation being observed at pH 8 and after addition of 60 mg/l of calcium. Based on PHREEQC-2 modelling, cadmium precipitates out as $Cd(OH)_2$ and $CdCO_3$. The precipitation of Cu is high when compared to that cadmium, probably because of its high concentration in the studied solution (4 mg/l for Cu and 100 µg/l of Cd). These concentrations were chosen based on the concentrations of Cd and Cu commonly found in groundwater.

[A]

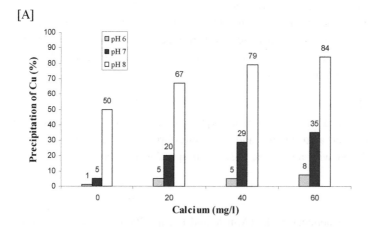

[B]

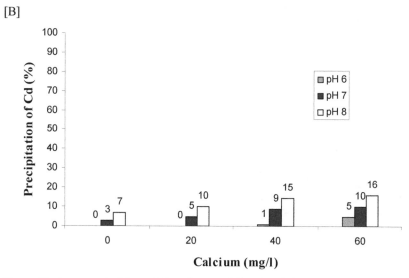

Figure 5.1: Assessment of the stability of copper [A] and cadmium [B]. Model water: Initial [Cu^{2+}] = 4 mg/l, [Cd^{2+}] = 100 µg/l, [HCO$_3^-$] = 100 mg/l, pH 6, 7 and 8, Ca 0-60 mg/l, shaking speed = 100 rpm,

5.3.2 Batch adsorption tests

Results from batch adsorption experiments (Figure 5.2) demonstrated that both IOCS and GFH could remove Cd^{2+} from aqueous solution. However, IOCS showed higher removal efficiencies. Increased Ca^{2+} concentrations in the model water at pH 6, 7 and 8 resulted in reduced Cd^{2+} removal efficiencies with both IOCS and GFH. For example, at pH 7, 33% of the Cd^{2+} was removed trough GFH and without Ca^{2+} addition. The removal decreased up to 6% when 60 mg/l of Ca was added to the model water. For IOCS and at pH 7; 81% and 63% of Cd were removed before and after addition of 60 mg/l of Ca, respectively. In addition, it was observed that the precipitation increased as the concentration of Ca^{2+} in the model water increased. Adsorption was found to be the predominant cadmium removal mechanism, except when GFH was used as adsorbent in combination with high calcium concentrations (60 mg/l) and high pH (7 or 8).

Similar results were obtained when Ca^{2+} was added to Cu^{2+} containing model water. This suggests that Ca^{2+} was competing with Cd^{2+} or Cu^{2+} for the surface adsorption sites of IOCS and GFH. It was also observed that Ca^{2+} presence in the model water increased precipitation of Cu^{2+} when IOCS was used as adsorbent, as predicted by the PHREEQC-2 modelling. Under the conditions applied, copper removal was predominantly achieved through adsorption when IOCS was used as adsorbent (except at pH 8 and in the presence of 40 and 60 mg/l of Ca^{2+}), while precipitation was the predominant mechanism when GFH was used. This is clearly seen on Figure 5.3 (A and B).

[A]

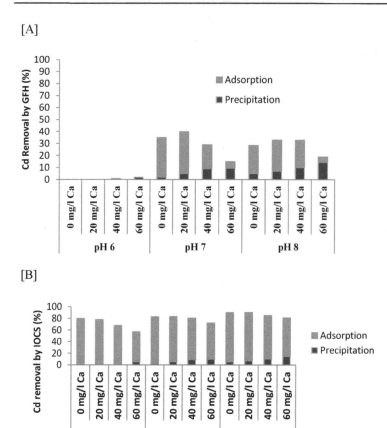

[B]

Figure 5.2. Effect of calcium on Cd removal by GFH [A] and IOCS [B], Model water: Initial $[Cd^{2+}]$ = 100 µg/l, $[HCO3-]$ = 100 mg/l, pH 6, 7 and 8, Ca 0-60 mg/l, adsorbent dosage = 0.1 g/l, shaking speed = 100 rpm,

[A]

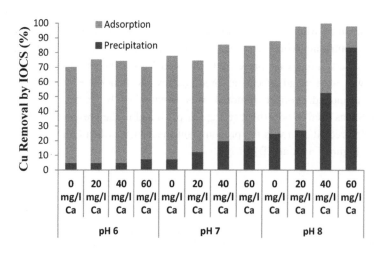

[B]

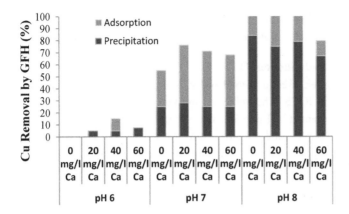

Figure 5.3: Effect of calcium on Cu^{2+} removal by IOCS [A] and GFH [B] Initial $[Cu^{2+}]$ = 4 mg/l, $[HCO_3^-]$ = 100 mg/l, adsorbent dosage = 0.1 g/l, shaking speed = 100 rpm,

The presence of calcium diminishes the number of adsorption sites of IOCS and GFH resulting in lower cadmium and copper removal. This is in agreement with the results obtained by Chen et al. (2002) and Zasoki and Burau, (1988).

5.3.3 Freundlich isotherm for cadmium removal

Batch adsorption experiments were conducted with varying IOCS dosages for getting isotherms of cadmium adsorption by IOCS at pH 6. Cd and pH 6 were chosen because they are the only parameters for which precipitation did not appear in the solution. Besides, IOCS showed better removal of Cd than GFH. The IOCS dosage varied between 0.05 and 0.6 g/l. Figure 5.4 shows the Freundlich isotherms of cadmium adsorption by IOCS.

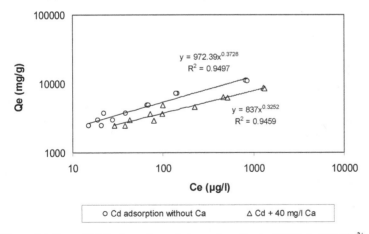

Figure 5.4: Freundlich isotherm for cadmium adsorption by IOCS at pH 6. $[Ca^{2+}]$ = 40 mg/l), $[HCO_3^-]$ = 20 mg/l and Initial $[Cd^{2+}]$ = 3 mg/l), IOCS dosages: 0.05 – 0.6 g/l and contact time 15 days

Figure 5.4 shows a good fit of the Freundlich model with high correlation coefficients (R^2 = 0.95) for Cd adsorption by IOCS with and without addition of calcium. The Freundlich isotherm constants (K) obtained are 972 (mg/g) (L/mg)$^{1/n}$ and 837 (mg/g) (L/mg)$^{1/n}$) for Cd adsorption without calcium and Cd adsorption with calcium addition, respectively.

5.3.4 Adsorption kinetics of Cadmium on IOCS

Figure 5.5 the pseudo-first order, pseudo second order, intraparticle diffusion and Elovich kinetic models of cadmium on IOCS. The values of the kinetic model parameters and the correlation coefficients are presented in Table 5.2. The equilibrium rate constants of pseudo-first order sorption (k_1) are 9.6 $\times 10^{-3}$ and 6.8 $\times 10^{-3}$ min^{-1} for Cd without and with calcium, respectively. The initial adsorption rate (h), the rate constants (k_2) and the correlation coefficients (R^2) for the pseudo second order sorption kinetic model were calculated and are also presented in Table 5.2.

The intraparticle diffusion rate constants k_{id} and the intercept were calculated from the slope of the plots. The intercepts of the plots are proportional to the thickness of the boundary layer (Acheampong et al., 2012) and are 0.48 and 1.29 mg.g^{-}1 for Cd without and with calcium, respectively. However, the correlation coefficients (R^2) for adsorption of Cd by IOCS were below 0.9.

In the Elovich model, the initial adsorption rate α and the desorption capacity β were obtained from the slope and the intercept, respectively. The correlation coefficients obtained indicate that the Elovich model fits the experimental data well, except for the adsorption of Cd by IOCS without calcium added to the model water.

Table 5.2: Effect of Ca^{2+} on adsorption kinetics parameters of cadmium sorption by IOCS

		Adsorption of Cd without Calcium	Adsorption of Cd + 40 mg/l Ca
First order Kinetic	k_1 (min^{-1})	9.6 $\times 10^{-3}$	6.8 $\times 10^{-3}$
	qe	3.1	1.6
	R^2	0.96	0.97
Pseudo Second order kinetic	K_2 (gmg^{-1}min^{-1})	1.180 $\times 10^{-3}$	1.47 $\times 10^{-2}$
	qe	0.3	0.17
	h (mgg^{-1}min^{-1})	1.87 $\times 10^{-2}$	4.93 $\times 10^{-2}$
	R^2	0.95	0.97
Intraparticle diffusion	K_{id} (mgg^{-1}min$^{-1/2}$)	0.16	0.13
	Intercept	0.48	1.29
	R^2	0.84	0.80
Elovich	α (mgg^{-1}min^{-1})		
	a (mgg^{-1}min^{-1})	2.09	0.95
	R^2	0.87	0.93

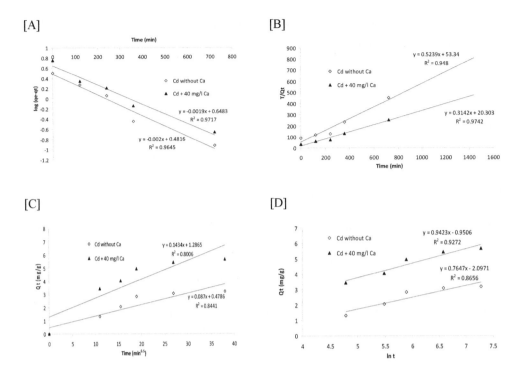

Figure 5.5: Pseudo-first order [A], Pseudo second order [B], Intraparticle diffusion [C] and Elovich kinetic models for cadmium adsorption by IOCS. Model water composition: [Cd] = 3 mg/l, [Ca^{2+}] = 0 and 40 mg/l, [HCO$_3^-$] = 20 mg/l, pH 6

5.4 Discussion

The results from short term batch experiments showed an inhibition of Cd and Cu sorption onto IOCS in the presence of calcium. The competitive effect of Ca and Cd on other adsobents than IOCS has been very well documented. This is for example reported by Escrig and Morell (1998) when analyzing the effect of Ca on soil adsorption of Cd and Zn. They found that a tenfold increase in the Ca concentration reduced the Cd adsorption capacity approximately by one third. The competition effect of Ca on Cd adsorption by sandy soil was also studied by Temminghoff et al. (2005). They found that, at an ionic strength of 0.03, Ca reduced Cd adsorption by 60–80%. Cd adsorption was also found to be higher in the presence of sodium compared to calcium due to the competition of Ca^{2+} ions with Cd^{2+} ions for adsorption sites on the surfaces of goethite (Mustafa et al. 2004). In the study conducted by Kiewiet and Wei-Chun (1991) on the effect of pH and calcium on lead and cadmium uptake by earthworms in water, it is suggested that Ca^{2+} and H$^+$ compete with lead and cadmium ions for binding sites of the earthworm biomass. The presence of background cations such as sodium, potassium or magnesium and anions such as chloride, nitrate, sulphate or acetate up to a concentration of 3.24 mmol/l was found to have no significant effect on the equilibrium uptake of Cd by brown, green and red seaweed (Hashim and Chu, 2004). However, the biosorbent uptake of Cd was markedly inhibited in the presence of Ca ions at 3.24 mmol/l.

The inhibition of cadmium adsorption also was confirmed by Benaissa and Benguella (2004) in studying the effect of cations on cadmium adsorption kinetics of chitin. Kang et al. (2010) also studied the influence of calcium precipitation on copper sorption induced by loosely bound extracellular polymeric substance (LB-EPS) from activated sludge. They found that, at lower Ca^{2+} concentrations ($\leq$ 80 mg/L), the Langmuir equation showed competitive adsorption between free Cu^{2+} and Ca^{2+} ions. However, at higher Ca^{2+} concentrations (> 80 mg/L), the competitive sorption between Cu^{2+} and Ca^{2+} was replaced by calcium precipitation. Ho and Ofomaja (2005) also studied the effects of calcium competition on lead sorption by palm kernel fibre. They found that the sorption capacity of lead and the initial sorption rate de-creased with an increase in calcium concentration and calcium appeared to enhance hydrogen release from the surface of the sorbent.

The adsorption of Ca, Cd and Cu on the surface of IOCS and GFH happens through electrostatic attraction at pH values above the pH_{PZC}. The better performance of IOCS in comparison to GFH can be related to the point of zero charge (pH_{PZC}) which is lower for IOCS compared to one of GFH. The charge on the surface of IOCS becomes negative staring from pH 7 while the surface charge of the GHF becomes negative staring from pH 8. Thus it will be easier for IOCS to attract cations at around pH 6 and 7 through electrostatic attraction than the GFH. Another explanation can be that the adsorption of Cu and Cd on IOCS is done through more than one layer (multilayer adsorption).

The Freundlich isotherm constants (K) obtained (972 and 837 (mg/g) (L/mg)$^{1/n}$) suggest that the adsorption capacity of IOCS was high. This is in agreement with results reported in the literature (Alduri 1996, Petrusevski et al. 2002). The highest adsorption capacity was observed for adsorption of Cd in the absence of calcium which confirms the results obtained in short batch experiments. Relatively low values of the coefficient $1/n$ <1 (0.33 and 0.37) suggests that changes in the equilibrium concentration of cadmium will have a limited effect on the IOCS cadmium adsorption capacities.

In comparison to the pseudo-first order kinetics, the pseudo-second order kinetics fits the experimental data better in terms of correlations coefficients (Figure 5.5[B]). Thus, one can conclude that the adsorption of Cd onto IOCS is likely a second order reaction with and without addition of calcium. The intraparticle diffusion plot (Figure 5.5) showed the correlation coefficients (R^2) for adsorption of Cd by IOCS below 0.9, which implies that the adsorption of Cd does not follow the intraparticle diffusion model. The bigger the intercept, the greater the contribution of the surface sorption to the rate controlling step, as stated by Acheampong et al. (2012). Hence, intraparticle diffusion is the main controlling process for Cd adsorption onto IOCS.

5.5 Conclusions

The knowledge of the effects of the groundwater quality matrix is essential to establish appropriate design parameters for adsorptive removal of heavy metals with iron oxide based adsorbents. From the results obtained in this study, the following conclusions can be drawn:

1. IOCS and GFH are capable of removing Cu and Cd from contaminated (ground)water, but IOCS removed Cu and Cd better than GFH.
2. Groundwater pH also is an important water quality parameter that can influence adsorption of Cu and Cd on IOCS and GFH. Better adsorption of Cd and Cu occurs at a low pH because higher pH values are favorable for precipitation.
3. The presence of calcium showed a competing effect with Cd^{2+} and Cu^{2+} for adsorption sites on IOCS and GFH. The Freundlich isotherm showed that the highest adsorption capacity was observed for adsorption of Cd in the absence of calcium, confirming the results obtained in short batch experiments.
4. The pseudo-second order kinetics fits the experimental data better than the pseudo-first order. Thus, the adsorption of Cd onto IOCS is likely a second-order reaction both with and without the addition of calcium.

5.6 References

Abdu-salam N., Adekola, F. A., 2005 The influence of pH and adsorbent concentration on adsorption of Lead, and Cadmium on natural goethite, *African Journal of Science and Technology (AJST)Science and Engineering Series Vol. 6, No. 2, pp. 55 – 66*

Acheampong M.A., Pereira J.P.C, Meulepas R.J.W and Lens. P.L.N, 2012, Kinetics modeling of Cu(II) biosorption on to cocunut shell and Moringa oleifera seeds from tropical regions, Environ. Techn. 33, 409-417

Amy, G., Chen, H.-W., Dinzo, A., Gunten, U., Brandhuber, P., Hund, R., Chowdhury, Z., Kommeni, S., Sinha, S., Jekel, M., and Banerjee, K. 2005. "Adsorbent Treatment Technologies for Arsenic Removal." *AWWA.*

Appelo C.A.J. and Postma D. 2005, Geochemistry, Groundwater and Pollution, 2nd edition, CRC Press, Taylor & Francis Group, 6000 Broken Sound Parkway NW, Suite 300

Benaissa H., Benguella B., (2004), Effect of anions and cations on cadmium sorption kinetics from aqueous solutions by chitin: experimental studies and modeling, Environmental pollution, vol 130 (2), pp 157-163

Chen, Z., Ma, W., Han, M., 2008. Biosorption of nickel and copper onto treated alga (*Undaria pinnatifida*): Application of isotherm and kinetic models. *Journal of Hazardous Materials*, 155(1-2):327-333

Chin-Hsing Lai, Chih-Yu Chen, Bai-Luh Wei; Shu- Hsing Yeh (2002). Cadmium adsorption on goethite-coated sand in the presence of humic acid. Water Res. 36, 4943 -4950

Escrig I., Morell I. (1998), Effect of Calcium on the Soil Adsorption of Cadmium and Zinc in Some Spanish Sandy Soils, Water, Air, and Soil Pollution, Volume 105, Issue 3-4, pp 507-520

Faust, S.D., Aly O.M., 1998, Chemistry of water Treatment, second edition, Lewis publisher, Boca Raton

Ghulam Mustafa, Balwant Singh, Rai S. Kookana (2004), Cadmium adsorption and desorption behaviour on goethite at low equilibrium concentrations: effects of pH and index cations, Chemosphere, Vol. 57, 10, pp 1325-1333

Hameed, B.H., 2008. Equilibrium and kinetic studies of methyl violet sorption by agricultural waste. *Journal of Hazardous Materials*, 154(1-3):204-212.

Hashim M.A. and Chu K.H. (2004), Biosorption of cadmium by brown, green, and red seaweeds, Chemical engineering journal, Vol 97 (2-3), pp 249-255

Ho Y.S., and McKay G, 2000 The Kinetic of Sorption of Divalent Metal ions on to Spagnum Moss Flat, Water Resource, 34(3), 735

Ho Y.S. and Wang C.C. 2004, Pseudo-isotherms for the sorption of cadmium ion onto tree fern, *Process Biochem* 39, 761-765

Ho Y.S. 2006, Review of second order models for adsorption systems. *J. Hazard. Mater.* 136, 681-689

Ho Y.S. and Ofomaja A. E., (2005), Effects of calcium competition on lead sorption by palm kernel fibre, Journal of Hazardous Materials 157–162

Huang, W.W., Wang, S.B., Zhu, Z.H., Li, L., Yao, X.D., Rudolph, V., Haghseresht, F., 2008. Phosphate removal from wastewater using red mud. *J. Hazard. Mater*, 158(1):35-42

Kang, Fuxing; Hamilton, Paul B.; Long, Jian; Wang, Qian 2010, Influence of calcium precipitation on copper sorption induced by loosely bound extracellular polymeric substance (LB-EPS) from activated sludge, Fundamental and applied liminology, 176, 173-181

Khaodhiar, S., Azizian, M.F., Osathaphan, K., Nelson, P.O., 2000. Copper, chromium, and arsenic adsorption and equilibrium modelling in an iron-oxide-coated sand, background electrolyte system. *Water Air Soil Poll* 119, 105–120

Kiewiet A.T and Wei-Chun Ma (1991), Effect of pH and calcium on lead and cadmium uptake by earthworms in water, Ecotoxicology and Environmental Safety, Vol 21, 1, pp 32-37

Lagergren, S., 1898. About the theory of so-called adsorption of soluble substances. *Kungliga Svenska Vetenskapsakademiens. Handlingar*, 24(4):1-39

Leyva-Ramos R.; Rangel-Mendez, J. R.; Men-doza-Barron, J.; Fuentes-Rubio, L.; Guer-rero-Coronado, R. M., (1997). Adsorption of cadmium (II) from aqueous solution on to activated carbon, Water Sci. Tech., 35 (7), 205-211

Pal B.N., 2001, Granular Ferric Hydroxide for Elimination of Arsenic from Drinking Water, *BUET-UNU International workshop on Technology for arsenic removal from drinking water*, 5-7 May, p 59-68

Petrusevski, B., Boere, J., Shahidullah, S.M., Sharma, S.K., Schippers, J.C., 2002, Adsorbent-based point-of-use system for arsenic removal in rural areas". *J.Water SRT-Aqua* 51, 135e144.

Rosa, S., Laranjeira, M.C.M., Riela, H.G., Fávere, V.T., 2008, Cross-linked quaternary chitosan
as an adsorbent for the removal of the reactive dye from aqueous solutions, *Journal of Hazardous Materials*, 155(1-2):253-260

Sharma S.K., Petrusevski B., Schippers J.C., 2002, Characterization of coated sand from iron removal plants. *J. Water Sci. Technol. Water Supply* 2.2, pp. 247-257.

Tan, I.A.W., Ahmad, A.L., Hameed, B.H., 2008. Adsorption of basic dye on high-surface-area activated carbon prepared from coconut husk: Equilibrium, kinetic and thermodynamic studies. *Journal of Hazardous Materials*, 154 (1-3):337-346.

Wan Ngah, W.S., Hanafiah, M.A.K.M., 2008. Adsorption of copper on rubber (*Hevea brasiliensis*) leaf powder: Kinetic, equilibrium and thermodynamic studies. *Biochemical Engineering Journal*, 39 (3):521-530.

Williams LE, Pittman JK, Hall JL.2000. *Emerging mechanisms for heavy metal transport in plants.* Biochimica et Biophysica Acta77803,1–23.

WHO 2004 Guidelines for Drinking water quality, Second edition Vol.2: Health Criteria and other supporting information. World health organization, Geneva, Switzland

Chapter 6: Effect of Fulvic Acid on Adsorptive Removal of Cr(VI) and As(V) from Groundwater by Iron Oxide Based Adsorbents

This Chapter is based on:

V. Uwamariya, B. Petrusevski, Y.M. Slokar, C. Aubry, P. N.L. Lens and G. Amy (2013), Effect of Fulvic Acid on Adsorptive Removal of Cr(VI) and As(V) from Groundwater by Iron Oxide Based Adsorbents, *Journal of Water, Air and Soil Pollution*, submitted

Abstract

In this study, ion oxide coated sand (IOCS) and granular ferric hydroxide (GFH) were used to study the effects of fulvic acid (FA) on the adsorptive removal of Cr(VI) and As(V). Characterization of IOCS and GFH by SEM/EDS as well as batch adsorption experiments were performed at different pH levels (6, 7 and 8). The surface of the virgin IOCS showed that Fe and O represent about 75% of the atomic composition and the presence of carbon was about 10%. The surface analysis of GFH showed that Fe and O represent, respectively, about 32% and 28% of the chemical composition. The adsorption tests with simultaneous presence of As(V) and FA on the one hand, and Cr(VI) with FA on the other hand, revealed that the role of FA was insignificant at all almost pH values with either IOCS or GFH as adsorbent. The influence of FA was only observed for the removal of As(V) and by IOCS or GFH at pH 6. It was also found out that organic matter (OM) was leaching out from the IOCS during experiments. The use of F-EEM revealed that humic-like, fulvic-like and protein-like organic matter fractions are present on the IOCS surface.

Key words: Adsorption, groundwater, As(V), Cr(VI), FA

6.1 Introduction

Groundwater is considered as a major source of drinking water in many countries because of its general advantages such as constant and good quality and accessibility. As a natural drinking water source, possible impurities present in groundwater are in most cases predictable. Impurities in groundwater originate from natural activities (non human), e.g., rain water, rock and soils in the earth crust, seawater intrusion, biological activities in the soil and animal activities. Natural contamination has become a challenging problem in drinking water production due to metal (e.g. arsenic, chromium, etc.) contamination of the groundwater sources throughout the world. This problem has increased the awareness of exposure of humans to these elements through drinking water. The World Health Organization (WHO, 2011) set a standard value for a maximum contaminant level (MCL) for arsenic and total chromium in drinking water at 10 μg/l and 50 μg/l, respectively.

As a naturally occurring element in the earth's crust, arsenic enters into aquifers through natural processes like mineral dissolution (e.g., pyrite oxidation), from geothermally influenced groundwater, or by reductive desorption and dissolution (Smedley and Kinniburgh, 2002). Arsenic occurs in groundwater predominantly in inorganic forms, with speciation and valence depending on oxidation-reduction conditions in the aquifer and the pH of the water. Generally, the reduced form of arsenic, As(III) - arsenite, is found in groundwater under anoxic conditions and the oxidized form, As(V) - arsenate, is generally found under oxic conditions. However both forms can be found in the same source. Depending on the pH, arsenate exists in four forms in aqueous solution (H_3AsO_4, $H_2AsO_4^-$, $HAsO_4^{2-}$, and AsO_4^{3-}) and arsenite exists in five forms ($H_4AsO_3^+$, H_3AsO_3, $H_2AsO_3^-$, $HAsO_3^{2-}$, and AsO_3^{3-}) (Wang *et al.* 2000}. In the pH range common for groundwater (6.5-

8.5), the predominant As(V) species are $H_2AsO_4^-$ and $HAsO_4^{2-}$, while As(III) is present as the neutral species H_3AsO_3 (Wang *et al.* 2000).

Chromium is a steel-grey metallic element also widely distributed in the earth's crust. Chromium is present in elevated concentrations in many groundwater sources used for drinking water in many countries around the world like the USA, Mexico, India, Canada, China, Scotland, Slovenia, Italy, Israel, etc. Naturally occurring chromium concentrations in groundwater are generally very low (less than 2 μg/l), although concentrations as high as 120 μg/l have been reported. In the aquatic environment, Cr is commonly found as hexavalent Cr(VI) and trivalent Cr(III). Cr(VI) compounds are much more soluble than Cr(III) and are much more toxic (mutagenic and carcinogenic) to microorganism, plants, animals and humans. In contrast, Cr(III) has relatively low toxicity and is immobile under moderately alkaline to slightly acidic conditions. Cr(VI) can cause liver and kidney damage, internal hemorrhage and respiratory disorders (Sharma, et al., 2008).

Various organic compounds generated by biological processes in nature, especially in the water environment, both in the water body (autochthonous material) and in the surrounding watershed (allochthonous material), are found in all surface and many groundwaters. These organic compounds are referred to as natural organic matter (NOM). They consist of a complex mixture of organic material like humic substances, carbohydrates, amino acids, carboxylic acids, proteins, hydrocarbons, etc (Croue, et al., 2000). In most places, the concentration of naturally occurring organic matter is low, and cannot be considered as contaminating natural water (Zaporozec, 2004). Aquatic natural organic matter (NOM) is present in all groundwater around the world, the concentration being between 0.5 to 10 mg/l of organic carbon. In general, NOM can be divided into two groups, the humic and non humic fractions. The humic fraction is more hydrophobic and comprises humic and fulvic acids. The non humic fraction is less hydrophobic and comprises hydrophilic acids, proteins, amino acids, and carbohydrates (Owen, et al., 1995). In most natural waters, humic substances usually dominate the NOM, contributing from 50 to > 90% of the dissolved organic carbon (DOC) which varies depending on the biological activities and seasonal cycles (Croue, et al., 2000). Humic molecules contain aromatic, carbonyl, carboxyl, methoxyl, and aliphatic units. Phenolic and carboxylic functional groups provide most of the protonation and metal complexation sites. Humic acid can be identified from its color which is dark brown to black.

Fulvic acid is the part of the humic substances that is soluble in water under all pH conditions; they remaining in solution after removal of humic acid by acidification. Differing from humic acid, fulvic acid is a light yellow to yellow-brown color. On the other hand, humin is the fraction of humic substances that is not soluble in water at any pH value which can be differentiated from its black color (Weber, 2009). Low concentrations of organics in natural waters make the determination of humic substances difficult, especially at neutral pH when both fulvic and humic acid are very soluble in water (Jucker et al., 1994). Fulvic acids are the most soluble group of organic matter and account for up to 90% of the dissolved humic material in natural waters. Fulvic acid contain aliphatic and aromatic components, especially carboxylic and phenolic functional groups, which provide protonation and complexation sites. The structure of fulvic acid contains not only a large proportion of aromatic moieties with HO⁻, HOOC⁻ and other oxygen rich groups, but also contain a portion of aliphatic groups. All of these functional groups may result in different hydrophobicity, aromaticity and polarity of fulvic acid. The variety of interaction taking place on mineral surfaces is largely determined by the differences in the amount and reactive ability of functional groups (Li et al., 2008).

Basically, NOM is considered non toxic, but its presence in groundwater gives an unwelcome color. Elevated color levels of groundwater could have an effect on taste and odor of water and can cause several problems during treatment and distribution of such groundwater. Another indirect, but considered very important effect is the formation of potentially harmful disinfection by-products and/or biological re-growth in the distribution systems (Genz et al., 2008).

It has been shown that metal (hydr) oxides play an important role in the adsorption and transport of organic substances in many natural aquatic systems. A considerable amount of work has been published on the binding of weak organic acids (Ali and Dzombak, 1996; Filius et al., 1997) and organic matter by minerals (Gu et al., 1994 and 1995; Wershaw et al., 1995). The results show that the adsorption of organic acids by mineral surfaces is dependent on pH and electrolyte concentration. The organic acids are bound over a large pH range, even at pH values well above the point of zero charge (PZC) of the adsorbing surface. Gu et al., (1995) investigated the binding of NOM by hematite using FTIR spectroscopy. They found that both carboxylic and hydroxylic groups are involved in the binding of NOM by hematite. Kaiser et al. (1997) showed similar results for organic matter binding by goethite. In order to provide further insight into the mechanisms and functional groups that are involved in the interactions, Gu et al. (1995) and Evanko and Dzombak (1998) studied the pH-dependent adsorption of organic acids containing carboxylic or phenolic groups. Compounds with carboxylic groups show an adsorption maximum at low pH, whereas compounds containing phenolic groups show a maximum at high pH. This suggests that carboxylic groups are relatively important in the binding of NOM at low pH, whereas hydroxyl groups are relatively important at high pH.

Several technologies have been used to remove heavy metals from ground water, including coagulation, precipitation, adsorption, ion exchange, membrane filtration, as well as in situ and biological processes (Pal 2001, Petrusevski et al. 2002, Faust and Aly, 1998). However, chemical precipitation and most of other methods become inefficient when heavy metals are present in trace concentrations. Adsorption is one of the few alternatives treatment technologies available for such situations.

Recent studies have shown that quartz sand and other filter media coated with iron, aluminium, manganese (hydro)oxides, or oxy-hydroxide are effective and inexpensive adsorbents for several metals (Khaodhiar et al. 2000, Petrusevski et al. 2002, Amy et al. 2005). Different factors including the nature and characteristics of the adsorbate, the ionic concentration, the presence of organic matter, the pH and the temperature can affect the effectiveness of the metal adsorption (Sharma et al. 2002, Abdu Salam and Adekola 2005).

The main goal of this research was to assess the effect of fulvic acid on the adsorptive removal of metal hydro-anions like arsenate -As(V)- and chromate -Cr(VI)- by IOCS and GFH at different pH values. The research focused on As(V) and Cr(VI) due to their toxicity and behaviour during adsorption processes such as charge and solubility of the species. FA was chosen as a model compound for NOM because FA is the most significant NOM acid component, and its molecules are highly soluble and relatively small. As several studies found that the negative charge of fulvic acid increases with increasing pH (Filius et al., 2000), fulvic acid is expected to compete with As(V) and Cr(VI) for adsorption surface of IOCS and GFH.

6.2 Materials and Experimental Methods

6.2.1 Adsorbents
GFH was obtained from the manufacturer GEH Wasserchemie in Osnabruck (Germany). IOCS was obtained from the Dutch Water Company Vitens, from the water treatment plant Brucht that treats groundwater with high iron content. For the screening and batch adsorption experiments, pulverized coating of IOCS and GFH ($< 63\mu m$) were used. After grinding, the bulk adsorbent was sieved to obtain the required fraction. The physical properties of the GFH and IOCS adsorbents are the same as the ones in Table 3.1.

6.2.2 Solutions preparation
A stock solution containing 100 mg/l of Cr(VI) was prepared from potassium dichromate (K_2CrO_7) while the stock solution of As(V) was prepared from a standard As(V) solution. The stock solution of FA was prepared from powdered Suwannee river fulvic acid from the International Humic Substances Society (IHSS). The required amount of sodium bicarbonate ($NaHCO_3$) was added to get the total concentration of 100 mg/l HCO_3^- in the model water.

6.2.3 Batch adsorption experiments
Several batch adsorption experiments were carried out to investigate the effect of FA on As(V) and Cr(VI) removal. FA concentration varied between 0 and 5 mg/l and the concentration of As(V) and Cr(VI) was 0.2 mg/l. Acid-cleaned and closed 500 ml plastic bottles, fitted with tubes for periodic sampling, were filled with synthetic water and the required amount of pulverized IOCS or GFH was added. Subsequently the pH was adjusted to the required values using 1M HNO$_3$ or NaOH solutions. Bottles were placed on an Innova 2100 rotary shaker at 100 rpm and kept at 20°C for 24 hours. Blank tests were carried out without adsorbent addition. All samples were filtered through a 0.45 μm membrane filter using a polypropylene syringe filter.

6.2.4 Analytical methods
As(V) and Cr(VI) were analyzed with an atomic absorption spectrometer (Thermo Elemental Solaar MQZe-GF 95) with an auto-sampler and a graphite furnace (AAS-GF). For dilution, acidified demineralised water was used. 5% Ni in 1 wt. % HNO$_3$ was used as matrix modifier. Water samples were passed through a 45μm filter and acidified with HCl to a pH below 2. IOSC and GFH were characterized by SEM/EDS methods (Modgi et al., 2006).

For NOM analysis, the total organic carbon (TOC) and dissolved organic carbon (DOC) concentrations were measured by a SHIMADZU TOC-VCPN analyzer. Fluorescence Excitation-Emission Matrix (EEM) was developed by using a Horiba Jobin Yvon FluoroMax-3 spectrofluorometer with a xenon lamp as the excitation source. Samples were filtered through a 0.45 um cellulose acetate membrane filter. The filters were soaked in Milli-Q water for 24 hours before use because the membrane filters could influence the reading by leaching DOC. Samples with concentrations of organic carbon above 20 mg/l were always diluted. The EEM contours were plotted in MATLAB. The EEM spectra, representing a 3-dimensional plot of fluorescence intensity versus excitation and emission wavelengths were used to reveal changes in protein-like organic matter (corresponding to an EEM peak at lower excitation/emission wavelengths) and humic-like organic matter (corresponding to an EEM peak at higher excitation/emission wavelengths).

The ultraviolet absorbance (UVA) and the specific ultraviolet absorbance (SUVA) measurements were performed using a UV-2501 PC spectrophotometer. Since water samples absorb UV light at wavelengths ranging from 200 to 300 nm, measurements of UV absorbance (UVA) were carried out at a wavelength of 254 nm. Specific UVA (SUVA) was then calculated by dividing a particular UVA_{254} by its corresponding DOC value. SUVA is an indicator of the relative aromaticity of organic components.

6.3. Results

6.3.1 SEM/EDS analysis of IOCS and GFH

SEM/EDS results (Figure 6.1A and B) showed that the surface of the virgin IOCS exhibits an accumulation of sub-micron sponge-like formations grains. C, O, Si, P, Ca and Fe were found in the IOCS coating. Iron and oxygen represent about 75% of the atomic composition and the presence of about 10% carbon was also observed. Si, P and Ca were detected as trace elements.

[A]

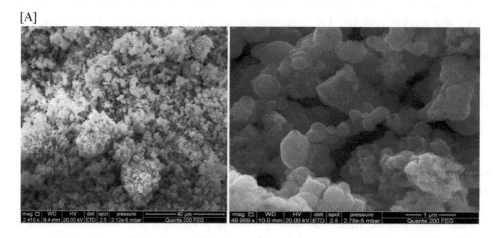

[B]

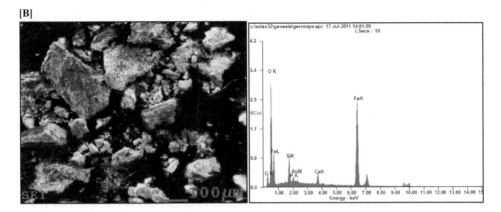

Figure 6.1: SEM view [A] and EDS results [B] of the virgin IOCS.

The surface analysis of the GFH (Figure 6.2) shows that Fe and O represent, respectively, about 32% and 28% of its chemical composition. Spot analyses indicate that the relative abundance of these elements onto the surface varies considerably. These spot analyzes also reveal traces of Na, Si, S, Cl, K and Ca. A particle rich in Al and O has been found on the surface.

[A]

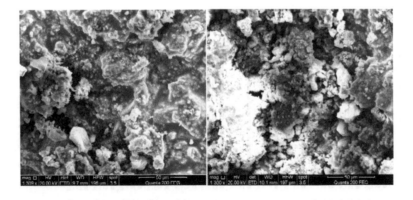

[B]

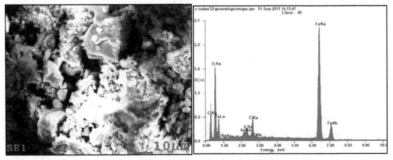

Figure 6.2: SEM view [A] and EDX results [B] of virgin GFH

6.3.2 Batch experiments

6.3.2.1 Stability of Cr(VI)

The stability of Cr(VI) was assessed in order to ensure that it was not precipitating under the conditions applied. In the absence and presence of FA, the stability of Cr(VI) was checked at pH 6, 7 and 8 for a period of 24 hours. The initial concentrations of Cr(VI) and FA were 0.2 mg/l and 2 mg/l, respectively.

It was found that Cr(VI) was stable under both conditions (with and without FA). Only a very slight reduction of 1% (in model water without FA) and 2% (in model water with FA) in concentration of Cr(VI) was observed at pH 6. The stability of As(V) was not tested because it is known that arsenic does not precipitate easily.

6.3.2.2 Screening tests on adsorption of As(V), Cr(VI) and FA by IOCS and GFH Separately

Adsorption tests with model waters containing As(V) or Cr(VI) and FA were performed at pH 6, 7 and 8. The efficiency of adsorption of IOCS and GFH are shown in Figure 6.3. Under the conditions applied, IOCS appears to be less effective in the As(V) removal compared to GFH. At pH 6, the adsorption of As(V) was 57% by IOCS versus 93% by GFH. Unlike the results of adsorption of As(V), Cr(VI) was adsorbed less effectively by both IOCS and GFH, but still the adsorption of Cr(VI) on GFH was much better than the adsorption on IOCS at pH 6 (39% and 8%, respectively).

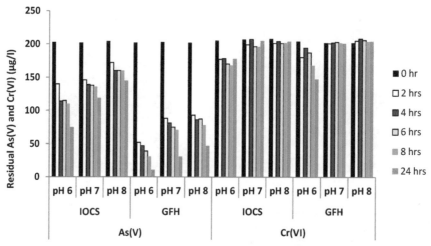

Figure 6.3: Adsorption of As(V) and Cr(VI) at pH 6, 7 and 8 by IOCS and GFH. Model water: Initial [As(V)] and [Cr(VI)] = 200 µg/l, [HCO$_3^-$] = 100 mg/l, dosage of adsorbent = 0.2 g/l. Shaker speed: 100 rpm.

The removal of FA was also assessed at the same pH values (6, 7 and 8). Figure 6.4 shows that FA is poorly adsorbed onto IOCS and GFH. As for As(V) and Cr(VI), FA was better adsorbed on GFH achieving 21%, 13% and 8% removal at pH 6, 7 and 8, respectively. In contrast, FA was released into the water instead of being removed by the IOCS: the increase was 5%, 12% and 6% at pH 6, 7 and 8, respectively.

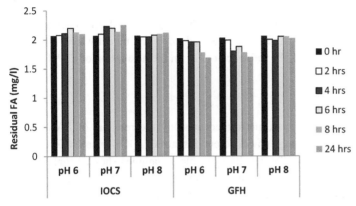

Figure 6.4: Adsorption of FA at pH 6, 7 and 8 by IOCS and GFH. Model water: Initial conc. of FA= 2 mg/l, [HCO$_3^-$] = 100 mg/l, dosage of adsorbent = 0.2 g/l. Shaker speed: 100 rpm.

6.3.2.3 Screening of adsorption of As(V) and Cr(VI) in the presence of FA by IOCS and GFH

The effect of FA on As(V) adsorption at pH 6 was assessed using model waters containing three concentrations of FA (0 mg/l, 2mg/l and 5 mg/l). A slight decrease in the removal efficiency of As(V) at increasing FA concentration was observed (Figure 6.5). As(V) removal efficiencies with IOCS of 63%, 50% and 40% were found for model water with 0 mg/l, 2 mg/l and 5 mg/l of FA, respectively, after 24 hours of contact time. However, the presence of FA did not affect the removal of As(V) by GFH.

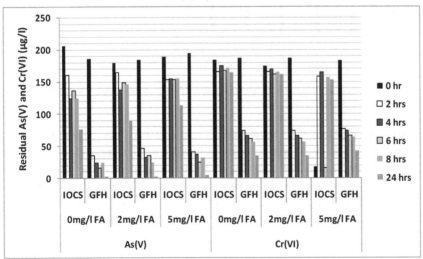

Figure 6.5: Removal of As(V) and Cr(VI) at pH 6 as a function of fulvic acid concentration. Model water: initial conc. of As(V) and Cr(VI) = 200 μg/l, HCO$_3^-$ = 100 mg/l, dosage of adsorbent = 0.2 g/l, conc. of FA = 0, 2, and 5 mg/l. Shaker speed: 100 rpm.

In the absence of FA, only 11% of the Cr(VI) was removed by the IOCS. The adsorption efficiency slightly decreased to 10% when 5 mg/l of FA were added to the model water.

Thus, FA did not appear to be a competitive species that can further reduce the poor performance of IOCS for Cr(VI) adsorption. GFH showed much better removal of Cr(VI). In the absence of FA, 82% of Cr(VI) was removed, while the removal efficiency reduced to 77% and 75% when 2 mg/l and 5 mg/l of FA were added to the model water, respectively. Thus, under the conditions applied, the influence of FA on the adsorption of Cr(VI) by IOCS and GFH was rather limited.

6.3.2.4 Removal of FA by IOCS

Additional batch adsorption tests were performed to study the removal of FA (present in model water at concentration of 2 mg/l) by different dosages of IOCS (blank, 0.2, 0.5, 1, 2, 5 and 10 g/l) at pH 6 (Figure 6.6). The results from blank samples showed that FA stayed relatively stable over the 10 days of the experiment. It was also observed that the concentration of FA in the model water increased with increasing adsorbent dosage. With prolonged contact time of 10 days, the concentration in FA increased by 59%, 122%, 238%, and 443% with IOCS dosages of 1, 2, 5 and 10 g/l, respectively. The results obtained strongly suggest that IOCS releases some OM into the model water (Figure 6.6).

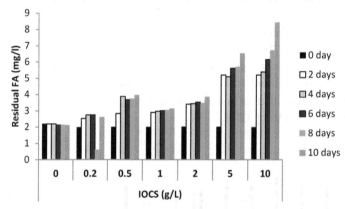

Figure 6.6: Adsorption tests of FA at pH 7 with higher dosages of IOCS. Model water: initial conc. of FA= 2 mg/l, [HCO$_3^-$] = 100 mg/l. Shaker speed: 100 rpm.

In order to verify the hypothesis of the release of organic matter from the IOCS, different amounts of IOCS (2, 5 and 10g//) were submerged into the demi-water, and shaken at 100 rpm for 96 hours. There was indeed a significant amount of organic matter leaching from IOCS into the demineralized water as clearly shown in Figure 6.7. The values of total organic carbon (TOC) and dissolved organic carbon (DOC) increase as the amount of IOCS increases.

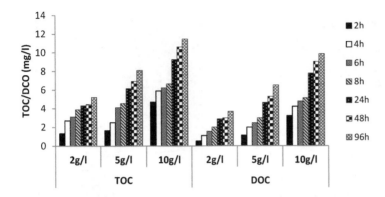

Figure 6.7: TOC and DOC that leached from IOCS in demineralized water after 96 hours of contact time

With an increase in pH, the concentration of organic matter increases as well (Table 6.1). As shown in Table 6.2, with an increase in IOCS dosage, the concentration of leached OM (as DOC) also increased, but the portion of leached amount of OM decreased. Results obtained indicate that the portion of OM leaching from IOCS was not linearly correlated with the total amount of OM introduced in the bottle with IOCS dosage.

Table 6.1: Leaching of organic matter (OM) from IOCS in the demineralized water

IOCS dosage (g/l)	Total amount of OM in the bottle (mg/l)	OM leached (mg/l)			% of OM leached		
		pH 6	pH 7	pH 8	pH 6	pH 7	pH 8
0.2	34	0.3	0.7	1.4	1.0	2.0	4.0
0.5	86	0.8	1.2	2.0	0.9	1.4	2.3
1	171	1.2	1.8	3.8	0.7	1.1	2.2
2	342	1.9	3.0	5.6	0.5	0.9	1.6
5	855	3.8	5.5	9.8	0.4	0.6	1.2
10	1710	4.6	8.0	13.0	0.3	0.5	0.8

Table 6.2: FEE-M analyses of OM leaching from IOCS in demineralized water at 100 rpm

	After 24 hours (Figure A)			After 96 Hours (Figure B)		
	Humic	Fulvic	Protein	Humic	Fulvic	Protein
Excitation	240	300	270	240	310	270
Emission	428	416	312	430	422	310
Intensity	1.608	1.397	1.062	2.352	2.515	1.26

[A]   [B]

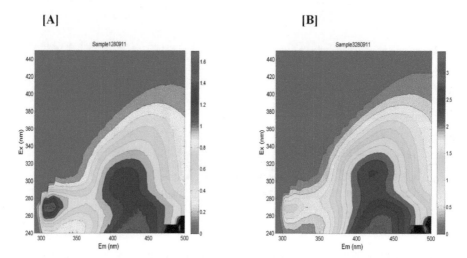

Figure 6.8: EEM spectra of OM that leached from IOCS in demineralized water [A] After 24 hours of contact time, [B] after 96 hours of contact time.

Different fractions of the organic matter leached from IOCS were analysed by Florescence Excitation-Emission Matrix (FEEM) as shown in Figure 6.8. Table 6.2 shows the excitation and emission of different fractions of OM. The excitation was observed at 240 nm, 300 nm and at 270 nm for humic-like, fulvic-like and protein-like organic matter fractions, respectively. The emission was observed at around 430 nm and 420 nm for humic-like and fulvic-like organic matter fractions, respectively, and at 310 nm for protein like organic matter fractions. Table 6.2 also shows that the amount of OM leached from IOCS increased with prolonged contact time. From 0 hr to 96 hours of contact time, the intensity increased by 46%, 80% and 19% for humic-like, fulvic-like and protein-like organic matter fractions, respectively. The increase in the intensity of the fulvic-like organic matter fraction of 80% clearly confirms the leaching of FA in significant amounts from IOCS.

6.3.2.5 Removal of NOM and As from Serbian groundwater by IOCS

In order to confirm that NOM released from IOCS affects the adsorption of As, an adsorption test was performed on groundwater from Zrenjanin (Serbia), that contains a very high DOC of 7.55 mg/l, and total As concentration of about 80 µg/L. Batch experiments with a contact time of 24 hours were performed at pH 6, pH 7 and pH 8. It was checked if NOM was removed during these adsorption tests, and consequently DOC and UVA were measured in samples taken for As analysis, and related SUVA values were calculated. Figure 6.9 shows that As contained in Serbian groundwater is well adsorbed by IOCS, with a removal efficiency of 78%, 65% and 37% at pH 6, 7 and 8, respectively. It was also found that the DOC concentration increased in line with results obtained in experiments with model water and IOCS as adsorbent.

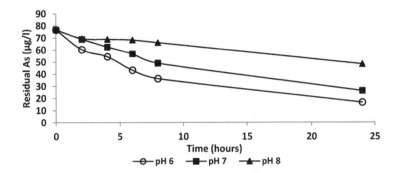

**Figure 6.9: Adsorptive removal of As from groundwater from Zrenjanin (Serbian) with initial [DOC] =
7.55 mg/L at different pH values. Working volume: 500 ml, IOCS dosage: 0.2 g/l, shaking speed: 100 rpm**

UVA and SUVA also were measured in As samples and showed that UVA results were
almost constant, while SUVA values were decreasing from 0.49 L/mg.m to 0.45 L/mg.m at
pH 6, and from 0.54 L/mg.m to 0.48 L/mg.m at pH 7, indicating that OM was released
instead of being adsorbed on the IOCS. However, the SUVA value changed only slightly at
pH 8 after 24 hours of contact time (Figure 6.10).

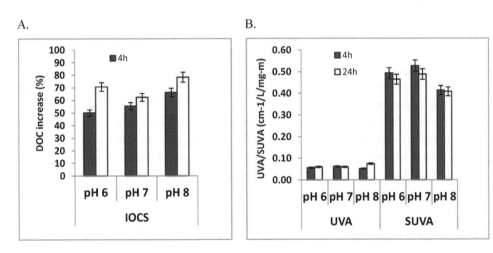

**Figure 6.10: Change in DOC concentration (A), UV254 absorbance and SUVA (B) during adsorption on
IOCS: Groundwater from Zrenjanin, Serbia (DOC =7.55 mg/L). IOCS dosage: 0.2 g/l, shaking speed:
100 rpm**

EEM of the Serbian groundwater was done to characterize the different fractions of OM
present. Serbian groundwater was found to contain humic-like and fulvic-like organic matter
fractions with high intensity compared to the ones leaching from the IOCS: 10.5 Int/mg C
and 5.7 Int/mg C for humic and fulvic fractions, respectively (Figure 6.11).

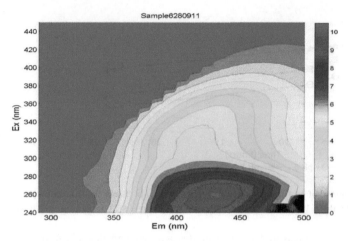

Figure 6.11: F-EEM spectra of OM contained in groundwater from Zrenjanin, Serbia

6.4 Discussion

6.4.1 Screening tests on adsorption of As(V), Cr(VI) and FA by IOCS and GFH Separately

This study showed that IOCS and GFH can remove both As(V) and Cr(VI). Their capability to remove As(V) and Cr(VI) can be related to the high amounts of iron and oxygen revealed by SEM/EDX analyses. However, the screening tests have shown that IOCS was less effective for As(V) adsorption as compared to GFH. This can be explained by the higher porosity and larger surface area of GFH, as shown in Table 3.1. Screening tests also showed that there was more effective adsorption of both As(V) and Cr(VI) by GFH and IOCS at pH 6. This is probably related to the point of zero charge (pH_{pzc}) of the IOCS and GFH (Table 3.1). As pH 6 is less than the pH_{pzc}, the surface of IOCS and GFH was positively charged and attracted stronger negatively species like Cr(VI) and As(V) (e.g. CrO_4^{2-} and $H_2AsO_4^-$).

The results from this study showed that FA was not at all adsorbed by IOCS and poorly adsorbed by GFH especially at higher pH value of 8. A limited FA removal was observed at pH 6 and pH 7 with 16% of FA adsorbed by GFH. This is in contrast with the results obtained by Gu et al. (1995) and Kaiser et al. (1997) who found that both carboxylic and hydroxylic groups are involved in the binding of NOM by hematite and goethite. However, a limited FA removal at low pH is in agreement with the results obtained by Evanko and Dzombak (1998) who found that organic acids containing carboxylic groups have a maximum adsorption at low pH, whereas compounds containing phenolic groups show a maximum adsorption at high pH. Yoon et al. (2004) also studied the adsorption of FA by boehmite (aluminium hydroxide) and reported that the adsorption of FA was better at lower pH and decreased as the pH was increased.

The variety of interactions taking place on mineral surfaces is largely determined by the differences in the amount and reactive ability of functional groups (Li et al. 2008). This can explain the low adsorption efficiency of FA by IOCS and GFH. In fact, the structure of FA contains not only a large proportion of aromatic moieties with HO⁻, HOOC⁻ and other oxygen rich groups, but also contain a portion of aliphatic groups. All these functional groups may result in different hydrophobicity, aromaticity and polarity of fulvic acid.

6.4.2 Screening of adsorption of As(V) and Cr(VI) in the presence of FA by IOCS and GFH

The results from the present study showed a slight decrease in the removal efficiency of As(V) at increasing FA concentration (Figure 6.5) and almost any effect on the removal efficiency of Cr(VI) by IOCS. The results also showed any effect of FA on the removal efficiency of As(V) and a negligible effect on the removal efficiency of Cr(VI) by GFH. Based on previous studies, several substances commonly found in natural water with arsenic sorption to HFO have been approved. These substances include bicarbonate (Appelo et al. 2002), silica (Swendlund and Webster) (1999), phosphate (Manning and Goldberg, 1996), and natural organic matter (NOM) (Redman et al. 2002). Beside Redman et al. (2002) showed that if NOM and As were incubated together with hematite, NOM dramatically delayed the attainment of sorption equilibrium and diminished the extent of sorption of both arsenate and arsenite. In terms of Cr(VI) removal, humic acid (HA) was found to exert an obvious inhibitory effect on Cr(VI) removal by Fe^0 nanoparticles. HA adsorbed on the surface of Fe^0 nano-particles and occupied the active surface sites, leading to the decrease in Cr(VI) reduction rates (Wang et al. 2010). Cr(VI) also forms no significant precipitates at levels encountered in potable water and does not strongly bind to natural organic matter.

Thus, FA is expected to act as a competitive ion for As(V) and Cr(VI) adsorption on IOCS and/or GFH, and can consequently reduce the adsorption capacity of the adsorbent. Alternatively, the presence of FA might create complexes and change the behavior of the species in the solution. Furthermore, negatively charged FA might influence the natural charge of the IOCS and GFH. Consequently, with the increasing concentration of FA species in the solution, negatively charged metals are expected to be less strongly attracted to more negatively charged adsorbents. However, under the condition applied in this study, the effect of FA on Cr(VI) and As(V) was negligible.

6.4.3 Removal of FA by IOCS

The results from batch adsorption tests on the removal of FA by different dosages of IOCS revealed that FA was released into the model water instead of being adsorbed (Figures 6.6 and 6.7). The release of OM into the model water by IOCS suggests that FA will likely not affect the removal of As(V) and Cr(VI) through competition for adsorption sites on IOCS, with an increase in the concentration of organic matter as the pH increases as well (Table 6.1) and the rate of leaching being slower with the higher OM concentration (IOCS dosage). This coincides with the observation made by Grybos et al. (2009) who found that any chemical process that increases the pH might play a role in the solubilization of the organic matter. As IOCS is a by-product from a water treatment plant, the release of OM from IOCS can be related to the quality of treated groundwater.

The non adsorption of FA by IOCS may probably due to the conditions applied in this study, such as pH for example. According to previous studies (Weng et al. 2009 and Chi and Amy, 2004), adsorption of natural organic matter was well observed at low pH and decreased as the pH was increased. Thus the pH range of natural groundwater (6.5-8.5) might not be favorable to the adsorption of FA and therefore not affect the adsorption of As(V) and Cr(V).

6.5 Conclusions

SEM/EDX analyses of IOCS and GFH showed a high iron and oxygen content, which suggests that IOCS and GFH can adsorb Cr(VI), As(V) and FA. However, the screening tests showed that IOCS was less effective in adsorption of As(V) and Cr(VI) as compared to GFH, likely due to differences in surface area and porosity of both adsorbents. FA was poorly adsorbed by IOCS and GFH, especially at higher pH values of 7 and 8.

FA was expected to act as a competitive ion of As(V) and Cr(VI) and can consequently reduce the adsorption capacity of the adsorbent. However, the influence of FA on the adsorption of As(V) and Cr(VI) by IOCS and GFH was very limited. Different functional groups may result in different hydrophobicity, aromaticity and polarity of fulvic acid. This might be the reason for low adsorption efficiency of FA by IOCS and GFH.

6.6 References

Abdu-salam N., Adekola, F. A., (2005) The influence of pH and adsorbent concentration on adsorption of Lead, and Cadmium on natural goethite, *African Journal of Science and Technology (AJST)Science and Engineering Series Vol. 6, No. 2, pp. 55 – 66*

Ali M. A. and Dzombak D. A. (1996) Competitive sorption of simple organic acids and sulfate on goethite. *Environ. Sci. Technol.* **26**, 2357–2364.

Amy, G., Chen, H.-W., Dinzo, A., Gunten, U., Brandhuber, P., Hund, R., Chowdhury, Z., Kommeni, S., Sinha, S., Jekel, M., and Banerjee, K. (2005). "Adsorbent Treatment Technologies for Arsenic Removal," *AWWA*.

Appelo C. A. J., Van der Weiden M. J. J., Tournassat C., and Charlet L. (2002), Surface complexation of ferrous iron and carbonate on ferrihydrite and the mobilization of arsenic: Environmental Science & Technology, 36, 3096-3103.

Chi F.H.and Amy G. L. (2004), Kinetic study on the sorption of dissolved natural organic matter onto different aquifer materials: the effects of hydrophobicity and functional groups, Journal of Colloid and Interface Science 274(2):380-91

Croue JP, Korshin GV, Benjamin M (2000) Characterization of natural organic matter in drinking water. AWWA Research Foundation and American Water Works Association

Evanko C. R. and Dzombak D. A. (1998) Influence of structural features on sorption of NOM- analogue organic acids to goethite. *Environ. Sci. Technol.* **32**, 2846–2855.

Faust, S.D., Aly O.M., (1998), Chemistry of water Treatment, second edition, Lewis publisher, Boca Raton

Filius J. D., Hiemstra T., and Van Riemsdijk W. H. (1997) Adsorption of small weak organic acids on goethite: Modeling of mechanisms. *J. Colloid Interface Sci.* 195, 368–380

Genz A, Baumgarten B, Goernitz M, Jekel M (2008) NOM removal by adsorption onto granular ferric hydroxide: Equilibrium, kinetics, filter and regeneration studies. Water Research 42: 238-248

Grybos M, Davranche M, Gruau G, Petitjean P, Pédrot M (2009) Increasing pH drives organic matter solubilization from wetland soils under reducing conditions. Geoderma 154: 13-19

Gu B., Schmitt J., Chem Z., Liang L., and McCarthy J. F. (1994) Adsorption and desorption of natural organic matter on iron oxide: Mechanisms and models. *Environ. Sci. Technol.* 28, 38–46.

Gu B., Schmitt J., Chem Z., Liang L., and McCarthy J. F. (1995) Adsorption and desorption of different organic matter fractions on iron oxide. *Geochim. Cosmochim. Acta* 59, 219–229.

Jucker C, Clark MM (1994) Adsorption of aquatic humic substances on hydrophobic ultrafiltration membranes. Journal of Membrane Science 97: 37-52

Kaiser K., Guggenberger G., Haumaier L., and Zech W. (1997) Dissolved organic matter sorption on subsoils and minerals studied by 13C-NMR and DRIFT spectroscopy. *Europ. J. Soil Sci.* **48**, 301–310.

Khaodhiar, S., Azizian, M.F., Osathaphan, K., Nelson, P.O., 2000, Copper, chromium, and arsenic adsorption and equilibrium modelling in an iron-oxide-coated sand, background electrolyte system. *Water Air Soil Poll* 119, 105–120.

Li A, Xu M, Li W, Wang X, Dai J (2008) Adsorption characterizations of fulvic acid fractions onto kaolinite. Journal of Environmental Sciences 20: 528-535

Manning B. A., and Goldberg S. (1996), Modeling competitive adsorption of arsenate with phosphate and molybdate on oxide minerals. Soil Sci. Soc. Am. J. 60, 121-131.

Modgi S., McQuaid M. E. and Englezos P. (2006), SEM/EDX analysis of Z-direction distribution of mineral content in paper along the cross-direction, Pulp and Paper Canada, 107:5, 124-127

Owen DM, Amy GL, Chowdhury ZK, Paode R, McCoy G, Viscosil K (1995) NOM characterization and treatability. AWWA January

Pal B.N., 2001, Granular Ferric Hydroxide for Elimination of Arsenic from Drinking Water, *BUET-UNU International workshop on Technology for arsenic removal from drinking water,* 5-7 May, p 59-68

Petrusevski, B., Boere, J., Shahidullah, S.M., Sharma, S.K., Schippers, J.C., (2002), Adsorbent-based point-of-use system for arsenic removal in rural areas". *J.Water SRT-Aqua* 51, 135-144

Redman A.D., Macalady D.L., and Ahmann D. (2002), Natural organic matter affects arsenic speciation and sorption onto hematite. *Environmental Science & Technology (36*13): 2889-2896.

Sharma S.K., Petrusevski B., Schippers J.C., (2002), Characterization of coated sand from iron removal plants. *J. Water Sci. Technol. Water Supply* 2.2, pp. 247-257.
Sharma SK, Petrusevski B, Gary A (2008) Chromium removal from water: a review. Journal of Water Supply: research and Technology 57: 541-553

Sharma SK, Petrusevski B, Gary A (2008) Chromium removal from water: a review. Journal of Water Supply: research and Technology 57: 541-553

Smedley P.L., Kinniburgh D.G., (2002) A review of the source, behaviour and distribution of arsenic in natural waters, Applied Geochemistry 17, 517–568

Swendlund P. J., and Webster J. G. (1999). Adsorption and polymerisation of silicic acid on ferrihydrite, and its effect on arsenic adsorption. Water Research 33, 3414-3422.

Wang J., Zhao F.J., Meharg A.A ., Raab A., Feldmann J., McGrath S.P., (2002) Mechanisms of arsenic hyperaccumulation in *Pteris vittata*. Uptake kinetics, interactions with phosphate, and arsenic speciation. Plant Physiology 130, 1552–1561

Wang Q., Cissoko N., Zhou M., Xu X. (2010), Effects and mechanism of humic acid on chromium (VI) removal by zero-valent iron (Fe^0) nanoparticles, Physics and Chemistry of the Earth, article in press

Weng L., Van Riemsdiik W.H., Hiemstra T. 2009, Effects of fulvic and humic acids on arsenate adsorption to goethite: experiments and modeling. Environ Sci Technol., 1;43(19):7198-7204

Wershaw R. L., Leenheer J. A., Sperline R. P., Song Y., Noll L. A., Melvin R. L., Rigatti G. P. (1995) Mechanisms of formation of humus coatings on mineral surfaces 3. Composition of adsorbed organic acids from compost leachate on alumina. *Colloids Surf.* 96, 93–104.

WHO 2004 Guidelines for Drinking water quality, Second edition Vol.2: Health Criteria and Other supporting information. World health organization, Geneva, Switzerland

Yoon TH, Johnson SB, Brown GE (2004) Adsorption of Suwannee River Fulvic Acid on Aluminum Oxyhydroxide Surfaces: An In Situ ATR-FTIR Study. Langmuir 20: 5655-5658 DOI 10.1021/la0499214

Zaporozec A. (2004), Groundwater contamination inventory, UNESCO IHP-VI, Series on Groundwater No. 2; (Contribution to IHP-V, Project 3.1), UNESCO, Paris.

http://www.ar.wroc.pl/~weber/humic.htm, accessed on the 28 June 2013

Chapter 7: Adsorption and surface complexation modelling of trace metal sorption onto iron oxide coated sand

Abstract

In this study, the removal of selected heavy metals, namely Cd(II), Cu(II), and Pb(II), by Iron Oxide Coated Sand (IOCS), a by-product from groundwater treatment plants, was screened. A series of batch adsorption experiments were conducted to study the effect of pH on the removal of Cd, Cu and Pb by IOCS. Single and combined metals experiments were performed in order to investigate the competition effects. XRF analysis showed that IOCS contains mainly hematite (Fe_2O_3) (approximately 85% of the total mass of minerals that could be identified by XRF). Chemical analysis revealed that the main constituent of IOCS is iron representing 32% on mass basis. Potentiometric mass titration (PMT) gave a value of pH of zero point charge of 7.0. Results for single metals using IOCS as an adsorbent showed that all metals included in the study can be very effectively removed with total removal efficiency as over 90% at all pH levels studied. The percentage of metals removed through precipitation was found to be metal specific: the highest for Cu (25%) and the lowest for Cd (2%) at pH 8. Concurrent presence of competing metals did not have a pronounced effect on the total metal removal efficiency, the range of reduction of total removal of Cu, Cd and Pb being between 1 and 4%. In terms of adsorption capacity, a competitive effect of metals was not observed except for Pb and Cu at pH 8 where the adsorption was decreased for 13% and 22%, respectively. Complexation modelling showed two type of complexes, one type associated with a weak site (Hfo_wOCd$^+$, Hfo_wOCu$^+$, Hfo_wOPb$^+$) and the other associated with a strong site (Hfo_sOCd$^+$, Hfo_sOCu$^+$, Hfo_sOPb$^+$) formed for all metals studied. Precipitation of Pb and Cu observed in batch experiments was confirmed in modelling from pH 6.75 and above. IOCS, being an inexpensive and available adsorbent can be used to treat water contaminated with heavy metals like Cd, Cu and Pb, however, pH is an important factor to be considered if one has to avoid precipitation, especially for the removal of Cu and Pb.

Key words: Modelling, Adsorption, Heavy metals, IOCS, Hydrous ferric oxide

7.1 Introduction

Heavy metals are often problematic environmental pollutants, with well-known toxic effects on living systems. By exceeding low acceptable concentration limits, as indicated by the WHO guidelines for drinking water (WHO, 2011), heavy metals are linked to human poisoning, and can cause learning disabilities, cancer and, at high concentrations even death. Heavy metals that can be occasionally found in ground water and some surface waters (e.g. urban run-off) typically include copper, nickel, cadmium, chromium, arsenic, lead and mercury (Wan Ngah and Hanafiah, 2008). The main sources of groundwater contamination in addition to natural pollution with heavy metals (e.g., in the case of arsenic), are seepage from disposal sites, industries such as plating, ceramics, glass, mining and battery manufacturing (Sharma and Al-Busaidi, 2001).

The most common technologies used for the removal of toxic metal ions from water are coagulation followed by separation, chemical precipitation, ion exchange, membrane separation (e.g. reverse osmosis), adsorption (e.g. activated carbon, activated alumina, iron oxides). However, chemical precipitation and other methods have been reported to be inefficient when contaminants are present in trace concentrations. In addition most of available physicochemical water treatment technologies are either too expensive or too technically complicated (Brown et al., 2000a). The process of adsorption is one of the few alternatives available (Huang, 1989). Metal removal through adsorption could be very effective, cheap, and simple to apply, assuming that low cost adsorbents with high adsorption capacity are available.

Previous studies have shown that sand and other filter media coated with iron, aluminium, or manganese oxide, hydroxide or oxyhydroxides are very good and inexpensive adsorbents (Sharma et al., 2002, Sharma et al., 1999). Iron Oxide Coated Sand (IOCS), a by-product from groundwater treatment plants has proved to be an effective adsorbent for the removal of iron (Sharma 1997) and arsenic (Petrusevski et al. 2002). Only limited results are available on the effectiveness of IOCS for the removal of other heavy metals. Insight in the effects of pH, concentration of metal, and presence of other metals is very limited.

Sorption reactions that take place at metal oxide surfaces can be fully described using surface complexation theory. The extent of adsorption or surface complexation depends on the type and density of the adsorption sites available and the nature of the adsorbing ion. Mineral surfaces, particularly metal (hydr)oxides, concentrate ions through both specific and non-specific complexation mechanisms. Specifically adsorbed ions are strongly surface-associated through covalent bonds formed by ligand exchange with surface hydroxyl groups (inner-sphere complexed), whereas non-specifically adsorbed ions are weakly surface-associated due to electrostatic interactions through an intervening water molecule (outer-sphere complexed) (Tadanier *et al.*, 2000, Stumm, 1992). Figure 1 illustrates a surface complexation formation of an ion on the hydrous oxide surface.

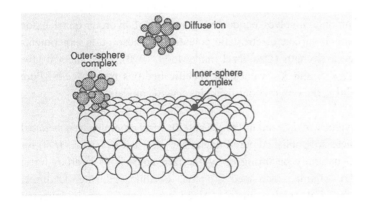

Figure 7.1: Surface complex formation of an ion on the hydrous oxide surface (Stumm and Morgan, 1981)

A simple method of distinguishing between inner-sphere and outer-sphere complexes is to assess the effect of ionic strength on the surface complex formation equilibria. A strong dependence of ionic strength is typical for an outer-sphere complex. Furthermore, outer-sphere complexes involve electrostatic bonding mechanisms and therefore are less stable than inner-sphere complexes, which involve appreciable covalent bonding along with ionic bonding.

Cation surface complexation by hydrous oxides involves formation of bonds with surface oxygen atoms and the release of protons from the surface, as presented by the following equations:

$$\equiv S-OH + M^{2+} \Leftrightarrow \equiv S-OM^{+} + H^{+} \tag{7.1}$$

or equivalently

$$\equiv S-OH + M^{2+} + H_2O \Leftrightarrow \equiv S-OMOH_2^{+} + H^{+} \tag{7.2}$$

where $\equiv$S-OH and M^{2+} represent a hydrated oxide surface and a divalent cation, respectively (Dzombak and Morel, 1990)

Anion adsorption by hydrous oxides occurs via ligand exchange reactions in which hydroxyl surface groups are replaced by the sorbing ions (Dzombak and Morel, 1990).

$$\equiv S-OH + A^{2-} + H^{+} \Leftrightarrow \equiv S-A^{-} + H_2O \tag{7.3}$$

and/or

$$\equiv S-OH + A^{2-} + 2H^{+} \rightarrow \equiv S-HA + H_2O \tag{7.4}$$

where A^{2-} represents a divalent anion.

Surface complexation (SC) theory has been used successfully to predict metal distribution in natural waters and sediments (*Wang et al.*, 1997, Smith, 1998, Davis *et al.*, 1998). Much of the SC research has investigated chemically simple systems in the laboratory. Surface complexation models use the law of mass-action, expressed as an equilibrium constant, to define protonation (K_{S+}), deprotonation (K_{S-}), and ion-specific sorption to a surface (K_{int}). The experimental procedure commonly involves conducting titrations of hydrous metal oxide suspensions to determine mineral surface electrostatic potential and adsorption experiments over a range of pH and ionic strength (Dzomback and Morel 1990). The intent of the experiments is to obtain K_{S-}, K_{S+}, and K_{int} values for specific hydrous metal oxides. Pure mineral K'_s are obtained by fitting the observations using inverse SC models.

Experimentally determined values of K'_s should be useable in more complex geochemical systems than those tested. There are published pure-mineral K_{S-}, K_{S+}, and K_{int} values (K'_s) for arsenate and arsenite SC on naturally occurring or synthesized amorphous hydrous ferric oxide (ferrihydrite or HFO), alumina, activated alumina, goethite, gibbsite, kaolinite, montmorillonite, illite, sulfides, soils, and sediments. A few researchers have investigated competition for sorption sites by binary (or more than 2) mixture of oxyanions or cations (Swedlund and Webster 1999). The study of surface complexation also has not advanced to the point where a ranking of anion and cation competition for hydrous metal oxide surfaces is available. The type and amount of each sorbing surface on a natural material must be known in order to use pure-mineral K'_s values to predict metal partitioning. Determining the surface composition and summing the parts, or measuring the bulk properties of the whole, are termed respectively the "component additivity" (CA) and the "generalized composite" (GC) methods (Davis *et al.*, 1998).

The electric layer geometries of the models differ, but they all reduce to mass action equations that are solved numerically. With respect to broad multicomponent modeling, the most important part of the commonality and formulation among the SC models is the universal application of the mass law constraint. The mass law formulation allows the sorption reactions to be used in the existing theoretical and mathematical framework of thermodynamic equilibrium models. Surface complexation reactions become a part of the general mathematical solution of an equilibrium state between an aqueous solution and a solid sorbing phase. This allows use of a single, broadly defined geochemical model to test several SC approaches (Gregory, 2001). The basic theory for surface-complexation reactions assumes that the number of active sites, T_s (eq), the specific area, A_s (m^2/g), and the mass, S_s (g), of the surface are known.

7.2 Material and methods

7.2.1 Adsorbent characterization
IOCS used in this study was obtained from the Dutch Water Supply Company Vitens, from the water treatment plant Brucht that treats groundwater with high iron concentration. IOCS was grounded to obtain a pulverized form with a size < 63 µm.

Mineralogy and chemical composition of IOCS
The mineralogy of IOCS was determined by semi quantitative X–Ray Fluorescence (XRF) spectrometry. The samples were pressed into powder tablets and the measurements were performed with a Philips PW 2400 WD–XRF spectrometer, the data evaluation was done with the Uniquant software.

The chemical composition of the IOCS was determined by coating extraction. Extraction was carried out by boiling a known amount of adsorbent in acidic solution. Different coating components were measured with: AAS-Flame Perkin-Elmer Analyst 200 (iron, manganese, magnesium, and calcium); spectrophotometer Perkin-Elmer Lambda 20 (phosphate); inductively coupled plasma (ICP) Perkin-Elmer Optima 3000 (silica); and AAS-GF Solaar MQZ (arsenic).

The organic content was obtained after drying the IOCS at $70^\circ C$ (24hrs), $105^\circ C$ (over night) and at $520^\circ C$ (2 hrs). The mass of organics was calculated from the difference of mass of IOCS obtained at $105^\circ C$ and $520^\circ C$.

Point of zero charge
For potentiometric mass titration (PMT), a blank solution and suspensions with two different IOCS masses (0.5 g and 1 g) were prepared with constant ionic strength of 0.03 M KNO_3 and room temperature ($25 \pm 0.1^\circ C$). The suspension was equilibrated for 24 hrs and the titration of each suspension was performed under N_2 gas. 1 M KOH was added to deprotonate the surface sites of IOCS. The suspension was then titrated with the 0.1M HNO_3 solution. The pH value was recorded after each addition of the HNO_3 solution. The same procedure was

also followed for the blank solution. The common intersection point of the three titration curves (pH vs V_{HNO3} added) was identified as the pH_{PZC} (Bourikas et al., 2003).

7.2.2 Batch adsorption experiments

Synthetic water was prepared by mixing milli-Q water with the required amount of Cu^{2+}, Cd^{2+}, and Pb^{2+} from 100 mg/l stock solutions. For all three metals, a concentration of 10^{-4} mol/l was used in batch experiments. For both types of synthetic waters, 100 mg/l of $NaHCO_3$ was added to increase the buffering capacity. Subsequently the pH was adjusted to the required values using 1M HNO_3 or NaOH solutions.

Acid-cleaned and closed 500 ml plastic bottles, fitted with tubes for periodic sampling, were filled with synthetic water and 0.2g/l pulverized IOCS (< 63µm) was added. Bottles were placed on an Innova 2100 rotary shaker at 100 rpm and kept at 20 ±1°C. Blank tests were carried out without the adsorbent addition. All samples were filtered through a 0.45 µm membrane filter using a polypropylene syringe filter. The contact time for batch experiments was 24 hours.

Batch adsorption experiments were performed with synthetic water containing single metal and different combinations of three metals. All of the metals were analyzed with an atomic absorption spectrometer (Thermo Elemental Solaar MQZe-GF 95) with an auto-sampler and a graphite furnace used as detector (AAS-GF). For dilution acidified demineralised water was used. Nickel nitrate (50g Ni/l) was used as matrix modifier.

7.2.3 Adsorption modelling

In this study, the Dzomback and Morel (1990) database for surface complexation on hydrous ferric oxide available in PHREEQC-2 hydro geochemical model, was used. The Dzomback and Morel's model defines complexation reactions on two populations of sites of ferric hyrdoxide (Table 1), a strong site Hfo_sOH and a weak site Hfo_wOH, while the Gouy-Chapman double layer equation was used for defining the surface potential as a function of the surface charge and ionic strength (Appelo and Postma, 2009).

Cu, Cd and Pb sorption on hydrous ferric oxide was simulated assuming two types of sites (weak and strong) available on the oxide surface. Protons and cations compete for the two types of binding sites, and equilibrium is described by mass-action equations. Activities of the surface species depend on the potential at the surface, which is due to the development of surface charge. The variation in sorption of cations Cu^{2+}, Cd^{2+} and Pb^{2+} on hydrous ferric oxides as a function of pH was performed for metal concentrations of 10^{-4} M in 0.1 M, 0.01M and 0.001M sodium nitrate electrolyte.

For data input to PHREEQC, the chemical equation for the mole-balance and mass-action expressions and the log$K$ (equilibrium constant) and its temperature dependence of surface species were defined through the SURFACE_SPECIES data block. Surface master species or types of surface sites were defined with the SURFACE_MASTER_SPECIES data block. The identity of the surfaces and the number of equivalents of each site type, the composition of the surface, the specific surface area, and the mass of the surface were defined with the SURFACE data block. The SURFACE_MASTER_SPECIES data block in the default database files defines a binding site named "Hfo" (hydrous ferric oxides) with two binding sites. The name of a binding site is composed of a name for the surface, "Hfo", optionally followed by an underscore and a lowercase binding site designation, "Hfo_w" and "Hfo_s" for "weak" and "strong", respectively, in the database files. The properties of hydrous ferric oxides are presented in Table 7.1

Table 7.1: Properties of Hfo in the Dzomback and Morel's (1990) database and sorption reactions for various ions

Weak sites Hfo_w (mol/mol Fe)	Strong sites Hfo_s (mol/mol Fe)	Surface area (m²/g)	Mol. wt (g/mol)	pH$_{PZC}$
0.2	0.005	550	88.85	8.11

Central to the SC model approach is that protonation and dissociation reactions and ion-specific complexation constants are reversible and apply over a range of pH and ionic strength conditions. The equilibrium constants K_{S-} and K_{S+} are determined for protonation-deprotonation reactions at the oxide surface. The protonation reactions with the surface are described by the two steps of reversible process in equations 7.5 and 7.6:

$$SOH + H^+ \Leftrightarrow SOH_2^+ \qquad K_{S+} = \frac{[SOH_2^+]}{[SOH][H^+]} \exp\frac{(F\psi_0)}{(RT)} \qquad (7.5)$$

and

$$SOH \Leftrightarrow SO^- + H^+ \qquad K_{S-} = \frac{[SO^-][H^+]}{[SOH]} \exp\frac{(-F\psi_0)}{(RT)} \qquad (7.6)$$

where F is the Faraday constant ($9.65.10^{-4}$ coulomb/mole), ψo is the surface potential in volts, R is the universal gas constant, and T is the absolute temperature. This exponential electrostatic term appended to the standard form of the equilibrium mass-action equation is used to account for the change in surface potential because of the adsorption of the modeled ion (Gregory, 2001). The K_{S-} and K_{S+} constants allow the surface-sorbing properties to change with changing pH. Constants for specific sorbing ions that meet these constraints are referred to as "intrinsic constants" or K_{int}. In order to apply these models to SC, K_{int} for surface reactions must be known for each surface to be used, each sorbing ion, and each site defined on the surface.

The sorption reaction equations for various ions considered in this study are shown in Equations 7.7-7.12:

Protons on strong sites:

$$\text{Hfo\_sOH} + H^+ = \text{Hfo\_sOH}^{2+} \qquad \log K_{a1} = -7.29 \qquad (7.7)$$

$$Hfo\_sOH = Hfo\_sO^- + H^+ \qquad\qquad logK_{a2} = -8.93 \qquad\qquad (7.8)$$

Protons on weak sites:
$$Hfo\_wOH + H^+ = Hfo\_wOH^{2+} \qquad\qquad logK_{a1} = -7.29 \qquad\qquad (7.9)$$

$$Hfo\_wOH = Hfo\_wO^- + H^+ \qquad\qquad logK_{a2} = -8.93 \qquad\qquad (7.10)$$
Transition metals (Cu^{2+}, Cd^{2+}, and Pb^{2+})

$$Hfo\_sOH + M^{m+} = Hfo\_sOM^{(m-1)+} + H^+ \qquad log\_K_1 = -4.374 + 1.166logK_{MOH} \quad (7.11)$$

$$Hfo\_wOH + M^{m+} = Hfo\_wOM^{(m-1)+} + H^+ \qquad log\_K_2 = -7.893 + 1.299logK_{MOH} \quad (7.12)$$

K_1 and K_2 represents the equilibrium constants for the sorption of metal on strong and weak sites, respectively, and K_{MOH} represents the first hydrolysis constant of the metal in water.

The database included in PHREEQC-2 is not complete, but constants of other elements are estimated with *Linear Free Energy Relations* (LFER). These are based on the regression of optimized surface complexation *K's* and K_{MOH} for the first hydrolysis constant of the metal in water or the second dissociation constant of the acid anion.

7.3 Results and discussion

7.3.1 Adsorbent characterization

Mineralogy and chemical composition

XRF results suggest that the dominant mineral present in the IOCS coating is hematite, representing approximately 85% by weight of the coating that could be detected by XRF. Chemical composition of IOCS is given in Table 7.2. Composition of the IOCS coating is influenced by the quality of groundwater. Consequently iron was found to be the predominant IOCS constituent, representing approximately one third IOCS mass. In addition to iron, organic matter represented approximately 17% of IOCS mass. IOCS in addition, has approximately 2% of phosphate, and 1% of calcium. Quartz, sand on which the coating was formed, represented only 5.4% of the total IOCS mass. This suggests that IOCS is likely a good adsorbent of heavy metals (Shokes 1999, Babel 2003). Other components present in significant concentration were SiO_2 (8.55 %), CaO (3.78%) and P_2O_5 (1.99%). Other compounds were found to be present in trace concentrations.

Table 7.2: Chemical composition of IOCS

Element	As	Ca	Fe	Mg	Mn	Org. Matter	O	P	Quartz	Si	Not known
Concent. (% by weight)	0.07	1.23	32.5	0.03	0.16	17.10	18.25	1.95	5.42	0.22	23.10

Point of zero charge

The IOCS point of zero charge (pH_{PZC}) was found to be 6.9 based on mass titration MT (Stanić et al. 2011), and 7.0 based on potentiometric mass titration PMT (Figure 7.2). The determined pH_{pzc} value is in agreement with values reported in the literature. Cornell and

Schwertman, (2003) reported pH_{PZC} for IOCS of 7.0 and pH_{PZC} for hematite between 7.5 and 8.0.

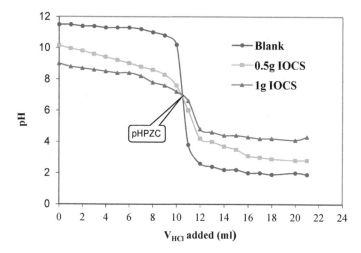

Figure 7.2: Point of zero charge of IOCS based on PMT: IOCS dosage: 0; 2.5 and 5 g/l; supporting electrolyte 0.03M KNO3, contact time: 24hrs, titrant solution: 0.1M HNO₃⁻, Temperature: 25± 0.1°C

7.3.2 Batch adsorption experiments

Single metal

Results from screening batch experiments for single metals, using IOCS as an adsorbent, showed that under the conditions applied, all metals included in the study can be very effectively removed (Figure 7.3). Total removal of Cu, Cd and Pb was more than 90% at all pH levels studied. However, some precipitation was observed for all studied metals; the precipitation increasing as the pH increases and being more pronounced at pH 8. Results of blanks (bottles without adsorbent) showed that precipitation, even at a relatively low pH of 6.0, significantly contributed to removal of Cu and Pb. Percentage of metals removed through precipitation was found to be metal specific: the highest for Cu (25%) and lowest for Cd (2%) at pH 8. In contrast to results obtained in batch adsorption experiments, precipitation of Cu and Pb at pH 6 in synthetic water with a single metal was not predicted by the PHRREQC-2 model, but precipitation of 2% and 5% of Cu and Pb, respectively, was observed.

Concurrent presence of 3 metals

In a natural water matrix, typically several metals are concurrently present. In order to assess possible competing effects of concurrently present metals, a separate set of experiments was conducted with synthetic water containing Cd, Cu and Pb (Figure 7.3B). Under conditions applied in the experiments, concurrent presence of competing metals did not have a pronounced effect on total metal removal efficiency. Due to the presence of competing ions, the range of reduction of total removal of Cu, Cd and Pb was between 1 and 4%. The reduction was 1% for Cd removal at all pH values studied, and for Cu removal at pH 6 and 8. It was 2% for Cu removal at pH 7 and Pb removal at pH 6 and 7. The reduction in total removal was 4% for only Pb at pH 8. With respect to precipitation, presence of competing metal ions increased precipitation of all metals studied. In general, the precipitation increased as the pH was increased and it was the highest for Pb and Cu at pH 8 with 29% and 26 %, respectively. In terms of adsorption capacity, no competitive effect of metals was observed except for Pb and Cu at pH 8 where the adsorption was decreased with 13% and 22%, respectively. Table 3 shows the difference in precipitation, adsorption and total removal of Cd, Cu and Pb when single and combined metals batch experiments are compared. The difference is made by taking the percentage of single adsorption minus the percentage obtained in combined adsorption tests.

[A]

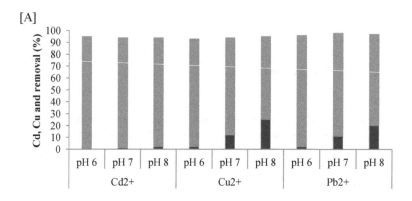

[B]

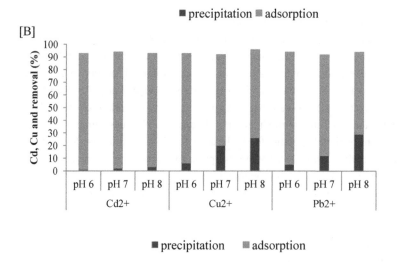

Figure 7.3: Single [A] and combined [B] removal of heavy metals in short☐term batch experiment. Synthetic water with initial metal concentration of 10^{-4}mole/L for Cd, Cu and Pb. c(HCO$_3^-$) = 100mg/l , Adsorbent dosage = 0.2 g/L, shaking speed = 100 rpm.

Table 7.3: Comparison of single and combined batch experiments for Cd, Cu and Pb removal by IOCS

		Difference in Precipitation	Difference in Adsorption	Difference in total removal
Cd^{2+}	pH 6	-1	6	1
	pH 7	-1	6	1
	pH 8	-1	0	1
Cu^{2+}	pH 6	-4	11	1
	pH 7	-8	10	2
	pH 8	-1	-22	1
Pb^{2+}	pH 6	-3	29	2
	pH 7	-1	17	2
	pH 8	-9	-13	4

As PHREEQC-2 is oriented toward system equilibrium, one can suggest that the systems used in batch adsorption experiments conducted in the laboratory were not equilibrated as we were operating at short term batch experiments.

7.3.3 Surface complexation modelling

Surface reactions are not half-reactions, so the master species is a physically real species and appears in mole-balance equations, and surface species may be anionic, cationic, or neutral. One has always to add the protonation reactions in Equations 7.7, 7.8, 7.9 and 7.10 to surface species formed between Hfo and cations. The change of the ionic strength did not influence the adsorption results in the modelling and only results for 0.1M $NaNO_3$ are herein presented.

Copper modelling

Besides the protonation reactions on strong and weak sites of the Hfo, the modelling of Cu adsorption on Hfo revealed two complexes, one on the weak site (Hfo_wOCu$^+$) and the other one on the strong site (Hfo_sOCu$^+$) as shown by equations 7 and 8.

$$Hfo\_sOH + Cu^{2+} = Hfo\_sOZn^+ + H^+ \qquad \log\_K = 2.89 \qquad (7.13)$$
$$Hfo\_wOH + Cu^{2+} = Hfo\_wOZn^+ + H^+ \quad \log\_K = 0.6 \qquad (7.14)$$

As Tadanier *et al.* (2000) and Stumm (1992) stipulated, outer-sphere complexes involve an electrostatic bonding mechanism and therefore are less stable than inner-sphere complexes, which involve appreciable covalent bonding along with ionic bonding. Hfo_sOCu$^+$ ions being strongly surface-associated through covalent bonds are considered as inner-sphere complexes whereas Hfo_wOCu$^+$ species are non-specifically adsorbed ions and are weakly surface-associated due to electrostatic interactions through an intervening water molecule. Thus, Hfo_wOCu$^+$ ions are considered as outer- sphere complexes. Figure 7.5 shows the distribution of Cu among the aqueous phase and strong and weak complexes.

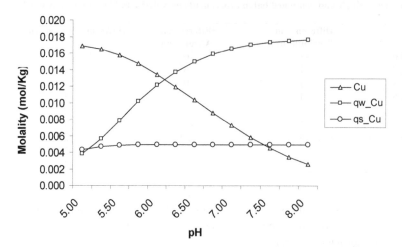

Figure 7.5: Distribution of copper among the aqueous phase and strong and weak surface sites of hydrous iron oxide as a function of pH for total copper concentrations of 10^{-4} mol/l

Beside the species found on the surface of hydrous ferric oxides (Hfo_sOCu$^+$, Hfo_sOH, Hfo_sO$^-$, Hfo_sOH^{2+}, Hfo_wOH, Hfo_wO$^-$, Hfo_wOCu$^+$ and Hfo_wOH^{+2}), other species are also formed in the solution. These species are Cu$^+$, Cu^{2+}, Cu$_2$(OH)$_2$$^{2+}$, CuOH$^+$, Cu(OH)$_2$, CuNO$_3$$^+$, Cu(NO$_3$)$_2$, Cu(OH)$_3$-, and Cu(OH)$_4$$^{-2}$. The modelling results show that copper starts precipitating out at pH 6.75 as tenorite (CuO). From pH 7.25, the precipitation is further accelerated by the formation of new precipitates of Cu(OH)$_2$ and Cu$_2$(OH)$_3$NO$_3$. This supports the results obtained in short-term batch experiments where precipitation of copper was even observed at pH 6. Table 7.4 shows different saturation indices for the formed precipitates.

Table 7.4: Variation of saturation indices with increase of pH

pH	6.75	7.0	7.25	7.5	7.75	8.0
CuO	0.69	1	1.28	1.54	1.76	1.95
Cu(OH)$_2$	-	-	0.25	0.50	0.73	0.92
Cu$_2$(OH)$_3$NO$_3$	-	-	0.24	0.50	0.70	084

Lead modelling
The modelling of Pb adsorption on Hfo revealed the following surface species:

Hfo_sOH + Pb^{+2} = Hfo_sOPb$^+$ + H$^+$	log_K 4.65	(7.15)
Hfo_wOH + H$^+$ = Hfo_wOH^{2+}	log_K 7.29	(7.16)
Hfo_wOH = Hfo_wO$^-$ + H$^+$	log_K -8.93	(7.17)
Hfo_wOH + Pb^{+2} = Hfo_wOPb$^+$ + H$^+$	log_K 0.32	(7.18)

Strong species formed at the surface of Hfo are Hfo_sOH^{2+}, Hfo_sO$^-$ and Hfo_sOPb$^+$, while weak species are Hfo_wOH^{2+}, Hfo_wO$^-$ and Hfo_wOPb$^+$. Other species found in the solution are $PbOH^+$, Pb^{+2}, $PbNO_3^+$, $Pb(NO_3)_2$, $Pb(OH)_2$, $Pb(OH)_3^-$, Pb_2OH^{+3}, $Pb(OH)_4^{-2}$, $Pb_3(OH)_4^{+2}$ and $Pb_4(OH)_4^{+4}$.

The modelling process was performed from pH = 5 to pH = 8 with 0.25 pH unit intervals. Starting from pH 6.75, precipitation occurred in the solution. This is indicated by the positive saturation index of $Pb(OH)_2$ as shown in Table 7.5. This has an impact on the adsorption process as most of the Pb^{2+} in solution starts precipitating at a pH value below the point of zero charge when the maximum adsorption is expected above pH 7 (pH_{pzc}). Thus, the precipitation of the Pb^{2+} is more pronounced and its adsorption is less important above pH_{pzc}. Like for copper, Hfo_sOPb^+ species are considered as inner-sphere complexes as they are strongly surface-associated through covalent bonds while Hfo_wOPb^+ species are considered as outer-sphere complexes and weakly surface-associated. However the adsorption of both complexes decreases as the pH increases, especially at pH values above 6.5. Figure 7.6 shows the variation of strong and weak species of Pb formed as well as the total concentration of Pb in the solution.

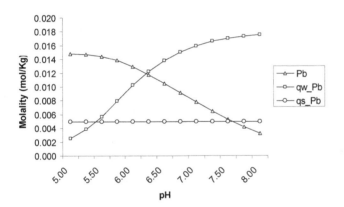

Figure 7.6: Distribution of lead among the aqueous phase and strong and weak surface sites of hydrous iron oxide as a function of pH for total lead concentrations of 10^{-4} M.

Table 7.5: Variation of saturation indices with increase of pH

pH	6.75	7.0	7.25	7.50	7.75	8.0
SI	0.19	0.53	0.85	1.14	1.40	1.63

Cadmium modelling
According to the PHREEQC-2 modelling different complexes are formed in the solution according to the equations:

$Hfo\_sOH + Cd^{+2} = Hfo\_sOCd^+ + H^+$	$log\_K = 0.47$	(7.19)
$Hfo\_wOH + H^+ = Hfo\_wOH^{2+}$	$log\_K = 7.29$	(7.20)
$Hfo\_wOH = Hfo\_wO^- + H^+$	$log\_K = -8.93$	(7.21)
$Hfo\_wOH + Cd^{+2} = Hfo\_wOCd^+ + H^+$	$log\_K = -2.91$	(7.22)

Species formed in the solution containing cadmium ions are Cd^{2+}, $CdNO_3^+$, $CdOH^+$, $Cd(NO_3)_2$, $Cd(OH)_2$, Cd_2OH^{3+}, $Cd(OH)_3^-$, $Cd(OH)_4^{2-}$. Hfo_s species are Hfo_sOH, Hfo_sOH^{2+}, Hfo_sO$^-$, Hfo_sOCd$^+$ and Hfo_w species are Hfo_wOH, Hfo_wOH^{2+}, Hfo_wO$^-$, and Hfo_wOCd$^+$.

Cd(II) typically forms inner-sphere complexes upon sorption onto various metal oxides (Boily et al., 2001; Papelis, 1995; Randall et al., 1999; Spadini et al., 1994). Spadini et al. (1994) conducted a series of X-ray absorption spectroscopy (XAS) experiments for Cd(II) on ferrihydrite for a range of surface loading (µmoles sorbate/m^2 sorbent) and pH values, and suggested that Cd(II) was predominantly adsorbed onto high affinity sites and that the Cd(II) coordination structure remained the same regardless of the surface coverage. Like for Cu and Pb, Hfo_sOCd$^+$ species are considered as inner-sphere complexes as they are strongly surface-associated through covalent bonds while Hfo_wOCd$^+$ are considered as outer-sphere complexes and weakly surface-associated species. The distribution of cadmium among the aqueous phase as well as strong and weak complexes are presented on Figure 7.7. As one can see, the decrease in total concentration of Cd is low in comparison with Cu and Pb. This confirms what was observed in batch experiments where the precipitation of Cu and Pb occurred above pH 6.5.

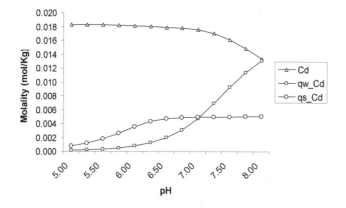

Figure 7.7: Distribution of cadmium among the aqueous phase and strong and weak surface sites of hydrous iron oxide as a function of pH for total cadmium concentrations of 10^{-4} M.

Figure 7.8 shows the adsorption edges of Cu, Cd and Pb. As it can be clearly seen, Pb is mostly adsorbed below pH 5, while Cd and Cu are well adsorbed below pH 6. From this, one can conclude that the maximum adsorption of all cations is better if the pH is lowered to less than 6. This will avoid the precipitation of Pb and Cu which starts at pH 6.75. The same conclusion can explain the high precipitation of Pb and Cu during batch experiments (Figures 7.3 and 7.4) at pH 7 and 8.

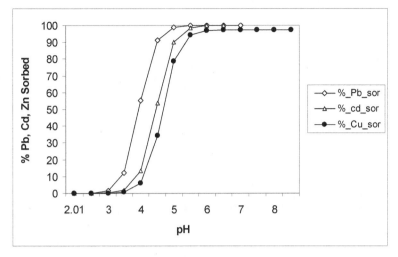

Figure 7.8: pH dependence of Cd, Cu and Pb in single solute modelling. Metal concentration 10^{-4} M and ionic strength of 0.1 M NaNO$_3$

The adsorption edge for Cu obtained in this study is in agreement with the Dzombak and Morel (1990) description who performed the adsorption of Cu on Hfo using a 2-site double layer surface complexation model (DLM) assuming that Cu adsorption occurs as a monodentate complex only on the strong site, according to Equation 23. This was later confirmed by Tracy et al. (2008) who measured Cu adsorption on HFO as a function of pH, ionic strength and sorbate/sorbent ratio.

$$Fe_{(s)}OH + Cu^{+2}_{(aq)} = >Fe_{(s)}OCu^{+} + H^{+}_{(aq)} \qquad\qquad (7.23)$$

Modelling of three cation mixture: cadmium, copper and lead
In order to assess the possible competing effect of concurrently present metals, modelling was conducted assuming water containing Cd, Cu and Pb together. A species distribution diagram is presented in Figure 7.9. Predominant species at pH below 6 are Cd^{2+}, Cu^{2+}, Pb^{2+} but also strong complexes of Pb and Cu (inner-sphere complexes are present). Strong complexes of Cd start to form into the solution above pH 6. Weak complexes are minor into the solution for all metals.

In contrast with the result obtained in batch experiments where the precipitation increased, no precipitation was predicted by PHREEQC-2 modelling from pH 4 to pH 9. In the equilibrium phase, $Cd(OH)_2$ and $Pb(OH)_2$ are formed but have negative saturation indices all along the pH range.

According to Susan et al. (2009) and Srivastava (2005) there should be competition of Cd and Pb for >FeOH sites and displacement of Cd from hydroxyl sites to ion exchange sites in the presence of Pb. This conclusion may also concern Cu as they are cations. Thus, the ion exchange must also be considered when multiple cations are present in contaminated water.

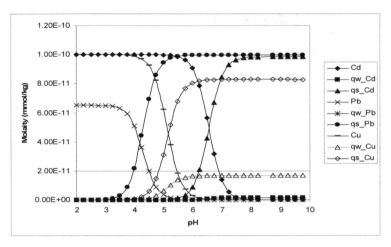

Figure 7.9: Species distribution diagramme of Cd, Cu and Pb modeled together in PHREEQC-2 for total Cd, Cu and Pb concentrations of 10^{-4} M.

7.4 Conclusions

The aim of this study was to screen the potential of IOCS to remove selected heavy metals under different experimental conditions, and describe the sorption reactions of Cd, Cu and Pb that take place at ferric oxide surfaces using surface complexation theory and modelling. Based on the results emerging from the study, the following conclusions are drawn:

- XRF analysis showed that IOCS coating contains manly hematite. Destruction of IOCS also showed that iron is the predominant element present in IOCS coating. XRD and the coating composition suggest that IOCS is likely a good adsorbent of heavy metals. IOCS pH_{PZC} established with PMT was around 7.0, in agreement with values reported in the literature for IOCS and hematite.
- Results from screening batch experiments for single metals, using IOCS as an adsorbent, showed that under conditions applied, all metals included in the study can be very effectively removed with total removal of Cu, Cd and Pb of more than 90% at all pH levels studied. Percentage of metals removed through precipitation was found to be metal specific: highest for Cu (25%) and lowest for Cd (2%) at pH 8. Under the conditions applied in the experiments, the concurrent presence of competing metals did not have a pronounced effect on the total metal removal efficiency. Due to the presence of competing ions, the range of reduction of total removal of Cu, Cd and Pb was between 1 and 4%. In terms of adsorption, competitive effects of metals were not observed except for Pb and Cu at pH 8.
- In complexation modelling, besides the protonation reactions on strong and weak sites of the Hfo, two type of complexes, one on the weak site (Hfo_wOCd^+, Hfo_wOCu^+, Hfo_wOPb^+) and the other one on the strong site (Hfo_sOCd^+, Hfo_sOCu^+, Hfo_sOPb^+) are formed for all of the metals studied. Cd, Cu and Pb mainly form inner sphere complexes as their formation does not depend on the ionic strength.

- IOCS being a cheap and available adsorbent, can be used to treat (ground)water contaminated with heavy metals like Cd, Cu and Pb, however, pH is an important factor to be considered if one has to avoid precipitation, especially for the removal of Cu and Pb.

7.5 References

Appelo C.A.J and Postma D. (2009) Geochemistry, groundwater and pollution, CRC press, 2nd edition

Babel S. and Kurniawan T.A., (2003), Low-cost adsorbents for heavy metals uptake from contaminated water: a review, *Journ Hazardous Material*, 97 (1-3), pp 219-243

Boily J.-F., Lutzenkirchen J., Balmes O., Beattie J. and Sjoberg S. (2001) Modelling proton binding at the goethite (a-FeOOH)-water interface. Coll. Surf. A 179, 11–27.

Brown P., Atly Jefcoat I., Parrish D., Gill S., Graham E., (2000a) Evaluation of the adsorptive capacity of peanut hull pellets for heavy metals in solution. Advances in Environmental Research 4: 19-29

Bourikas K., Vakros J., Kordulis C., Lycourghiotis A. (2003) Potentiometric mass titrations: experimental and theoretical establishment of a new technique for determining the point of zero charge (PZC) for metal (Hydr)oxides. J Phys Chem 107:9441–9451

Cornell R.M., Schwertmann U. (2003), The Iron oxides, Structure, Properties, Reactions, Occurrences and Use, Wiley-VCH, 2nd edition, p237

Davis J.A., J.A. Coston, D.B. Kent, and C.C. Fuller (1998), Application of the surface complexation concept to complex mineral assemblages, *Environ. Sci. Technol.* 32(19):2820-2828.

Dzombak D.A. and Morel F.M.M., (1990), Surface Complexation Modeling: Hydrous Ferric Oxide. John Wylie and Sons, New York.

Gregory P.M, (2001), Surface complexation modeling of arsenic in natural water and sediment systems, PhD thesis, New Mexico Institute of Mining and Technology Department of Earth and Environmental Science

Huang P.M., (1989), Rates of Soil Chemistry Processes, Sparks D.L., Suarez D.L. Eds.; *Soil Science Society of America* : Madison WI, pp. 191-230.

Papelis,C.1995.X-ray photoelectron spectroscopic studies of cadmium and selenite adsorption on aluminum oxides. Environ. Sci.Technol. 29:1526–1533

Petrusevski, B., Boere, J., Shahidullah, S.M., Sharma, S.K., Schippers, J.C., (2002), Adsorbent-based point-of-use system for arsenic removal in rural areas". *J.Water SRT-Aqua* 51, 135e144

Sharma R.S. and Al-Busaidi T.S. (2001). Groundwater pollution due to a tailings Dam. Eng. Geol., 60: 235 –244.

Randall S. R., Sherman D. M., Ragnarsdottir K. V., and Collins C. R. (1999) The mechanism of cadmium surface complexation on iron oxyhydroxide minerals. Geochim. Cosmochim. Acta 63, 2971–2987

Sharma R.S. and Al-Busaidi T.S. (2001). Groundwater pollution due to a tailings Dam. Eng. Geol., 60: 235 –244.

Sharma S.K., Greetham M., Schippers J.C., (1999), Adsorption of iron (II) onto filter media,

J. Water SRT Aqua 48 (3), pp. 84-91

Sharma S.K., Petrusevski B., Schippers J.C., (2002), Characterization of coated sand from iron removal plants. *J. Water Sci. Technol. Water Supply* 2.2, pp. 247-257.

Smith S.L., Jaffe, P.R., (1998), Modeling the Transport and Reactions of Trace Metal in Water Saturated Soils and Sediments. *Water Res. Research*, 34(11):3135-3147.

Stanić, N., Uwamariya, V., Slokar, Y.M. and Petruševski, B. (2011). "Competitive removal of selected heavy metals commonly found in urban stormwater runoff ". *In preparation*

Shokes T. E. and Möller G. (1999), Removal of Dissolved Heavy Metals from Acid Rock Drainage Using Iron Metal, *Environ. Sci. Technol.*, **33** (2), pp 282–287

Spadini L., Manceau A., Schindler P. W., and Charlet L. (1994) Structure and stability of Cd surface complexes on ferric oxides. 1. Results from EXAFS spectroscopy. J. Colloid Interface Sci.168, 73–86.

Srivastava P., Singh B. and Angove M. (2005) Competitive adsorption behavior of heavy metals on kaolinite, J. Colloid Interface Sci. 290, 28–38.

Stumm W.; (1992), Chemistry of the solid-water interface, John Wiley & Sons, Inc.,

Swdlund P.J., Webster J.G., (1999), Adsorption and polymerization of silicic acid on ferrihydrite and its effect on Arsenic Adsorption, *Wat. Res.* 33(16):3413-3422

Tadanier C. J. and Eick M. J., (2000) Surface Complexation Modeling of Pi-Goethite Adsorption Behavior: Integrating Adsorption Edge and Surface Charging Data. V. M. Goldschmidt Conference-Oxford, U. K.

Tracy J.L., Carla M. K., Christopher J. L., Melinda S. S. and Soumya D. (2008), Surface complexation modeling of Cu(II) adsorption on mixtures of hydrous ferric oxide and kaolinite, *Geochemical Transactions,* 9:9

Wang, F. Chen, J., Forsling, W., (1997), Modeling Sorption of Trace Metals on Natural Sediments by Surface Complexation Model. *Environ. Sci. Technol.* 31:448-453

Wan Ngah, W.S. and Hanafiah, M.A.K.M., (2008), Removal of heavy metal ions from wastewater by chemically modified plant wastes as adsorbents: A Review. Journal of Bioresource Technology 99(2008). 3935-3948.

WHO (2011), Guidelines for Drinking water quality,: Health Criteria and other supporting information. *World health organization*, Geneva, Switzerland Second edition Vol. 2

Chapter 8: Summary and conclusions

8.1. Introduction

Groundwater is rain water or water from surface water bodies, like lakes or streams that soaks into the soil and bedrock and is stored underground in the tiny spaces between rocks and particles of soil. In general groundwater is preferred as a source of drinking water because of its convenient availability and its constant and good quality. However this source is vulnerable to contamination by several substances. Substances that can pollute groundwater are divided into substances that occur naturally and substances produced or introduced by human activities (Schippers *et al.,* 2007). Naturally-occurring substances causing pollution of groundwater include for example, iron, manganese, ammonium, fluoride, methane arsenic, and radionuclides. Substances resulting from human activities include, for example, nitrates, pesticides, synthetic organic chemicals and hydrocarbons, heavy metals, etc. Because of their increasing discharge and toxic nature, the presence of heavy metals in groundwater is of big interest. High concentrations of heavy metals in water are undesirables because of adverse effects on health, environmental toxicity, corrosion of pipes and the aesthetic quality of the water. From this point of view there is a need for treating groundwater contaminated by heavy metals.

In Rwanda, surface water is the main source of drinking water and groundwater represents only 10 % of the total drinking water produced. However the distribution of drinking water is still inadequate and clearly groundwater is needed to supplement surface water sources. Groundwater in Rwanda also remains unexplored field and very limited information is available on the quality of this source of drinking water. Thus there is a need of screening the quality of the groundwater in Rwanda.

8.2 Needs for research, goal and objectives

As acceptable quality limits relative to micropollutant contents in drinking water are becoming increasingly lower, more efficient elimination treatment processes are required. The most common technologies used in the removal of toxic metal ions from water are the process of coagulation, precipitation, ion exchange, membranes separation by reverse osmosis, adsorption onto coagulant flocs and adsorption onto sorptive media (e.g. activated carbon, activated alumina, iron oxides, etc). It has been shown that chemical precipitation and other methods become inefficient when contaminants are present in trace concentrations. Chemical precipitation has several limitations: i) precipitation is often ineffective if the metals are complexed or if they are presents as anions, ii) the lowest metal concentration achievable is limited by the solubility product, iii) those metals which do precipitate may form small particles that do not settle readily (Benjamin et al, 1996), iv) the process generates large volume of sludge which requires disposal and is normally hazardous due to high concentration of heavy metals. The main disadvantages of reverse osmosis are the high initial investment and operational cost and they waste a lot of water (Faust and Aly, 1998) while the disadvantages of ion exchange process are: i) high cost, ii) the spent regenerate must be disposed off, iii) difficult to remove the metals that are present in uncharged form. The process of adsorption is one of the few alternatives available for such situations (Afssa, 2005,

Huang 1989). It is also known that selective adsorption retain elements that conventional treatments are unable to eliminate (Benjamin and Leckie, 1981).

Recent studies have shown that sand and other filter media coated with iron, aluminium, or manganese oxide, hydroxide or oxihydroxide were very good and inexpensive adsorbents which, in some cases, are more effective than the methods usually employed, such as precipitation-coprecipitation or adsorption on activated carbon grains (Sharma et al., 2002, Sharma et al., 1999, Benjamin and Leckie, 1981). This phenomenon was demonstrated after having observed that iron and manganese in particular were more effectively eliminated using old rapid sand filters than filters containing new sand. According to previous studies at UNESCO-IHE, Iron Oxide Coated Sand (IOCS) has been proved to be an effective adsorbent for the removal of heavy metals such as iron, arsenic and chromium (Sharma 1997, Petrusevski et al. 2000, Sunil 2004, Tessema 2004). The advantage of using IOCS is its high adsorptive capacity and the fact that the IOCS is a by-product of drinking water industry. It has also been reported that IOCS can effectively remove copper, nickel, cadmium and lead from groundwater (Tekeste, 2003). Granular ferric hydroxide (GFH) (commercially available) iron oxide based media is highly used for the removal of arsenic from drinking water. It is increasingly used as adsorbent for the removal of heavy metals from drinking and waste water. GFH has also shown the promising result as compared to activated alumina (Pal, 2001). IOCS and GFH have been tested in the removal of Cu, Cd, Cr(III) and Cr(VI) from urban storm water run-off (Devendra, 2007). The results show that IOCS is capable of removing all heavy metals studied and GFH is capable to remove Cu, and Cr but not Cd. However the competitive effect and the effect of other ions have not been studied. The effect of pH also needs further research.

The goal of this study was to improve the methods of adsorptive removal of selected metals present as cations or oxyanions (e.g. cadmium, cooper, arsenic and chromium) from groundwater by using different iron (hydro) oxides based media (iron oxide coated sand – IOCS and granular ferric hydroxide-GFH)). Furthermore, the study gave a better understanding on the interaction of a solute with a surface to be characterized in terms of the fundamental physical and chemical properties of the groundwater, the sorbent (IOCS and GFH) and the adsorbate (heavy metals).

The present study had the following objectives:
- To screen the quality of groundwater with focus on the presence of selected heavy metals in selected areas of Rwanda
- To characterize adsorbents used in this study namely IOCS, GFH with focus on their mineralogy
- To study the effect of organic matter, pH and selected interfering ions (Ca^{2+}, PO_4^{3-}) on the adsorption of selected heavy metals
- To give a better understanding on the mechanism involved in surface complexes formation and develop a Surface Complexation Model (SCM) that describes the processes involved in adsorptive removal of selected heavy metals by IOCS and GFH.

8.3 Screening of groundwater quality in Rwanda

The screening was done in the Eastern part of the country, especially in Nyagatare District. The District of Nyagatare experiences small quantities of rain and hot temperatures. Apart from the River Muvumba, Akagera and Umuyanja Rivers, there is no other consistent river that can be exploited by the population in Nyagatare. The weak river network constitutes a serious handicap to responding to the needs of water for people and animals. The total boreholes drilled in Nyagatare District are 91 but, in this study, groundwater samples were collected from 20 boreholes and 27 parameters (pH, temperature (T), turbidity, conductivity, dissolved oxygen (DO), total dissolved solids (TDS), total alkalinity (TA), total hardness (TH), Na, K, Ca, Mg, NH_3, NO_2^-, NO_3^-, F^-, Cl^-, SO_4^{2-}, PO_4^{3-}, Fe^{2+}, Mn^{2+} and Zn^{2+}) were analysed. However only 22 parameters are reported because the values of the other 5 parameters (Cu, Cr, Cd, Pb and As) were below the detection limits. The total hardness varied between 10 and 662 ppm, 7 samples fall under soft class, 3 samples fall under moderately hard class, 7 samples fall under hard and 3 samples fall under very hard class. The calculation of $\%Na^+$, RSC and SAR showed that Nyagatare groundwater is suitable for irrigation. Nyagatare District having abundant granite and *granite rocks* being igneous rocks, this can explain the source of fluoride found in groundwater. The source of EC, TDS, ammonia and nitrite in Nyagatare groundwater can be related to human activities by application of fertilizers and manures.

With a piper diagram representation, most of sampled sites are mainly sodium and potassium type and, for a few of them, there is no dominant type. In terms of anions, few sites have chloride groundwater type, one has bicarbonate groundwater type and others have no dominant anions.

PCA results showed that the extracted components represent the variables well. The extracted components explain nearly 94% of the variability in the original 22 variables, so that one can considerably reduce the complexity of the data set by using these components, with only a 6% loss of information. Six components were extracted. The first component was most highly correlated with fluoride, pH and sulfate, the second component was most highly correlated with calcium and total hardness and the third component is most highly correlated with total alkalinity. The fourth the fifth and the sixth components are mostly correlated with potassium, iron and magnesium, respectively.

8.4 Effect of water matrix on adsorptive removal of heavy metals

The knowledge of the effects of the groundwater quality matrix is essential to establish appropriate design parameters for adsorptive removal of heavy metals with iron oxide based adsorbents. From the results obtained in this study, the following conclusions can be drawn:
The presence of calcium in the model water enhanced As(III) adsorption on IOCS at all pH values studied (6, 7, 8) while the effect of calcium on As(V) removal by IOCS was limited at pH values 6 and 7 and strongly improved As(V) removal at pH 8. Similar results were obtained for As(III) and As(V) adsorption by GFH. The Freundlich isotherm constants (K) obtained (845-3357 $(mg/g)(L/mg)^{1/n}$) showed that the adsorption capacity of IOCS and GFH

was high and the highest K values were obtained when calcium was present in the model water. The kinetic study showed that adsorption of As(III) and As(V) onto GFH follows a second order reaction with and without addition of calcium while the adsorption of As(III) and As(V) onto IOCS follows a first-order reaction without calcium addition, and moves to the second reaction order kinetics when calcium is added. The main controlling process for As(III) adsorption is intraparticle diffusion while the surface adsorption contributes greatly to the adsorption of As(V). The RSSCT study confirmed the beneficial effect of calcium on As(III) and As(V) adsorption onto IOCS and GFH, which can significantly prolong the operational length of adsorptive filters.

The effect of calcium on cation adsorption by IOCS and GFH was also studied. IOCS and GFH showed the capability to remov Cu and Cd from contaminated (ground) water, but IOCS removed Cu and Cd better than GFH. Groundwater pH also is an important water quality parameter that can influence adsorption of Cu and Cd on IOCS and GFH. Better adsorption of Cd and Cu were obtained at low pH values because higher pH values are favorable for precipitation. The presence of calcium showed competition effect with Cd^{2+} and Cu^{2+} for adsorption sites on IOCS and GFH. The Freundlich isotherm showed that highest adsorption capacity was observed for adsorption of Cd in the absence of calcium which confirms the results obtained in short batch experiments. The pseudo-second order kinetics fits the experimental data better than the pseudo-first order in terms of correlations coefficients. Thus, one can conclude that the adsorption of Cd onto IOCS is likely a second-order reaction with and without addition of calcium.

The effect of PO_4^{3-} on Cr(VI), Cd^{2+} adsorption by IOCS and GFH was also assed. IOCS and GFH showed the potential to adsorb Cr(VI), Cd^{2+} and PO_4^{3-} through the combination of electrostatic and complexation mechanisms. GFH showed higher adsorption efficiency for both Cr(VI) and PO_4^{3-} than IOCS, but IOCS showed better removal of Cd^{2+}. For both adsorbents; better removal of Cr(VI), Cd^{2+} and PO_4^{3-} was observed at low pH value (pH 6). For Cd^{2+}, the precipitation was observed at higher pH values of 7 and 8.5, even predominating over adsorption at pH 8.5. Better adsorption of Cr(VI) and PO_4^{3-} at low pH value can be explained by electrostatic attraction forces due to the positively charged adsorbent surface. Presence of PO_4^{3-} decreased the adsorption capacities of Cr(VI) and vice versa. Cr(VI) and PO_4^{3-}, being both negatively charged and having both affinity for iron oxides, they compete for the surface adsorption sites on GFH and IOCS. Presence of PO_4^{3-} increased the adsorption capacities of Cd^{2+} by increasing the number of negative charges at the surface of the adsorbents. Thus, Cr(VI) removal by GFH and IOCS from contaminated groundwater is less favorable when PO_4^{3-} is present, while the opposite is observed when PO_4^{3-} is added to groundwater contaminated by Cd^{2+}.

The assessment on the effect of FA on the adsorption of As(V) and Cr(VI) was also performed. As a function of pH and time, the concentration of total Cr(VI) was found to be stable during 24 hrs of the experiment in the absence and presence of fulvic acid. Therefore, the influence of FA on the reduction of Cr(VI) to Cr(III) was not significant. In the adsorption tests with individual components, at pH 6, 7, and 8, As(V) removal by GFH was better than by IOCS. In the same range of pH, GFH performed slightly better than IOCS in the adsorptive removal of Cr(VI) and FA was poorly adsorbed by either, IOCS or GFH. A small removal was found at pH 6 by GFH. The negative charge of the metals at pH > 6.8 and the pH_{pzc} of the adsorbent appear to be the main reasons of this occurrence. In adsorption tests with the simultaneous presence of As(V) and FA, it was found that the effect of FA was almost insignificant at all pH values with either IOCS or GFH as adsorbent. An influence of FA was only detected in the removal of As(V) by IOCS GFH at pH 6. High As(V) removal (99%) occurred at pH 6 by GFH. In the adsorption experiments with simultaneous presence of Cr(VI) and FA by IOCS and GFH at pH 6, the role of FA as a competitive ion was insignificant with a poor performance of IOCS and GFH on the removal of Cr(VI). The highest Cr(VI) removal (82%) was provided by GFH. It was found that OM was leaching from IOCS during experiments. The use of EEM revealed that humic-like, fulvic-like and protein-like organic matter fractions were all presents on the IOCS. Tests performed on a Serbian groundwater showed the release of OM from IOCS with an increase of DOC and decrease of SUVA values when IOCS is used as adsorbent.

8.5 Surface complexation modelling

Sorption reactions that take place at metal oxide surfaces can be fully described using surface complexation theory. The extent of adsorption or surface complexation depends on the type and density of the adsorption sites available and the nature of the adsorbing ion. The aim of modelling surface complexation was to describe the sorption reactions of Cd, Cu and Pb that take place at ferric oxide surfaces using surface complexation theory. The modelling was done based on the laboratory results and the following conclusions are drawn: XRF analysis showed that IOCS coating contains manly hematite. Destruction of IOCS also showed that iron is the predominant element present in IOCS coating. XRD and the coating composition suggest that IOCS is likely a good adsorbent of heavy metals. IOCS pH_{PZC} established with PMT was around 7.0, in agreement with values reported in the literature for IOCS and hematite. Results from screening batch experiments for single metals, using IOCS as an adsorbent, showed that under conditions applied, all metals included in the study can be very effectively removed with total removal of Cu, Cd and Pb of more than 90% at all pH levels studied. Percentage of metals removed through precipitation was found to be metal specific: highest for Cu (25%) and lowest for Cd (2%) at pH 8. Under the conditions applied in the experiments, the concurrent presence of competing metals did not have a pronounced effect on the total metal removal efficiency. Due to the presence of competing ions, the range of reduction of total removal of Cu, Cd and Pb was between 1 and 4%. In terms of adsorption, competitive effects of metals were not observed except for Pb and Cu at pH 8. In surface complexation modelling, besides the protonation reactions on strong and weak sites of the Hfo, two type of complexes, one on the weak site (Hfo_wOCd$^+$, Hfo_wOCu$^+$, Hfo_wOPb$^+$) and the other one on the strong site (Hfo_sOCd$^+$, Hfo_sOCu$^+$, Hfo_sOPb$^+$) are formed for all of the metals studied. Cd, Cu and Pb mainly form inner sphere complexes as their formation does not depend on the ionic strength.

IOCS being a cheap and available adsorbent, it can be used to treat (ground)water contaminated with heavy metals like Cd, Cu and Pb. However, pH is an important factor to be considered if one has to avoid precipitation, especially for the removal of Cu and Pb.

8.6 Some recommendations for further studies

1. Due to financial limitations, only groundwater quality in Nyagatare District was screened. It is recommended to do a wide inventory for all existing wells and spring in Rwanda in order to have the overview of groundwater quality for the whole country.
2. The investigation of the effects of other parameters than pH, calcium, organics and phosphate on the adsorption of heavy metals by IOCS and GFH is highly recommended.
3. To run a pilot plant in order to asses the application of the results obtained from the laboratory is of high importance
4. The regeneration capability of IOCS and GFH is also recommended for further studies.

Samenvatting

In het algemeen heeft grondwater de voorkeur als bron van drinkwater vanwege de beschikbaarheid en de constante en goede kwaliteit. Grondwater is echter gevoelig voor verontreiniging door verscheidene stoffen. Stoffen die het grondwater kunnen verontreinigen zijn onderverdeeld in stoffen die van nature voorkomen en stoffen die worden geproduceerd of vrijkomen bij menselijke activiteiten. Natuurlijk voorkomende stoffen die tot verontreiniging van het grondwater kunnen leiden zijn bijvoorbeeld ijzer, mangaan, ammonium, fluoride, methaan, arseen en radionucliden. Stoffen ten gevolge van menselijke activiteiten omvatten, bijvoorbeeld, nitraten, pesticiden, synthetische organische chemicaliën en koolwaterstoffen, zware metalen, etc.

Aanvaardbare kwaliteitsnormen voor de concentraties aan microverontreinigingen in drinkwater worden steeds lager en efficiënte verwijderingprocessen worden geïmplementeerd om aan deze eisen te voldoen. Metaal verontreinigingen in een lage concentratie zijn moeilijk uit water te verwijderen. Chemische precipitatie en andere methoden worden inefficiënt wanneer de verontreinigingen aanwezig zijn in zeer lage concentraties. Het adsorptie proces is een van de weinige mogelijkheden voor dergelijke situaties. Recente studies hebben aangetoond dat zand of andere filtermedia bekleed met ijzer, aluminium of magnesium oxide, hydroxide of oxihydroxide zeer goede, goedkope adsorptiemiddelen zijn die in sommige gevallen effectiever zijn dan de gewoonlijk toegepaste methoden, zoals precipitatie-coprecipitatie of adsorptie aan actieve kool. Selectieve adsorptie kan ook elementen verwijderen die niet door de conventionele behandelingsmethoden worden verwijderd. Dit werd gedemonstreerd, nadat geconstateerd was dat in het bijzonder ijzer en mangaan effectiever verwijderd kunnen worden met behulp van oude filters dan door filters met nieuw zand. Dit kan, in de meeste gevallen, worden verklaard door een katalytische werking van de oxide afzettingen op het oppervlak van de zandkorrels.

In deze studie werd de adsorptie methode gebruikt voor de verwijdering van de geselecteerde zware metalen, aanwezig als kationen (Cd^{2+}, Cu^{2+} and Pb^{2+}) of oxyanionen (Cr(VI) en As(V)), door middel van ijzeroxide gecoat zand (IOCS) en granulair ijzerhydroxide (GFH). De effecten van de pH, natuurlijke organische stof (fulvozuren (FA)) en interfererende ionen (PO_4^{3-} en Ca^{2+}) op de adsorptie-efficiëntie werden ook onderzocht. Het oppervlaktecomplexatie model voor de adsorptie van Cd^{2+}, Cu^{2+} and Pb^{2+} werd bestudeerd teneinde de sorptie reacties die plaatsvinden op het oppervlak van het adsorptiemiddel te beschrijven. Batch-adsorptie testen en snelle kleinschalige kolomproeven (RSST) werden gebruikt als laboratorium methoden.

Rwanda maakt hoofdzakelijk gebruik van oppervlaktewater als bron van drinkwater en het gebruik van grondwater blijft een onontgonnen gebied met slechts zeer beperkte informatie over de kwaliteit van deze bron van drinkwater. In dit onderzoek werd de grondwaterkwaliteit gescreend in de oostelijke provincie (Nyagatare District), waar het grondwater de belangrijkste bron van drinkwater is. Voor de bepaling van de fysischchemische eigenschappen van het Nyagatare grondwater werden 27 waterkwaliteitsparameters geanalyseerd. De resultaten toonden dat de troebelheid en geleidbaarheid voor alle bemonsterde locaties binnen het bereik van acceptabele waarden voor drinkwater lagen. Onder de 20 bemonsterde locaties, hadden 12 locaties een pH-waarde binnen de drinkwater normen, 6 sites hebben zuur water en 2 sites alkalisch water. Voor alle bemonsterde plaatsen werd de opgeloste zuurstof concentratie te laag bevonden, wat aangeeft dat Nyagatare grondwater anoxisch is. De troebelheid is laag, behalve voor één locatie, en slechts vier sites zijn binnen het acceptabele bereik van de totale alkaliteit. De totale hardheid is groter dan de limieten voor 5 sites en de concentratie van belangrijke kationen (Ca^{2+}, Na^+, K^+ and Mg^{2+}) en de belangrijkste anionen (F^-, Cl^-, PO_4^{3-} and SO_4^{2-}) vallen binnen de drinkwaternormen voor alle bemonsterde sites. De ammoniakconcentratie was minder dan 3 mg/l op twee locaties. Ook de NO_2^- and NO_3^- concentraties respecteren de WHO (2011) richtwaarden (2 mg/l and 50 mg/l, respectievelijk). Wat betreft de concentratie van zware metalen hadden alle bemonsterde plaatsen Fe^{2+} waarden hoger dan 0.3 mg/l, wat de bovenste aanvaardbare concentratie is van de meeste nationale normen voor drinkwater (inclusief Rwanda), en tien plaatsen hadden Mn^{2+} concentraties hoger dan 0.1 mg/l, welke aanbevolen wordt door verschillende nationale normen om esthetische en operationele problemen te voorkomen. Wat andere zware metalen betreft, respecteert Zn^{2+} de norm voor alle bemonsterde sites, behalve voor alle plaatsen in Rwempasha en Rwimiyaga. Ondanks dat de belangrijkste focus van dit onderzoek het verwijderen van zware metalen is, blijken de concentraties van Pb^{2+}, Cd^{2+}, Cu^{2+}, As and Cr in het Nyagatare grondwater lager dan de detectielimieten te zijn.

Met de Piper diagram voorstelling vallen de meeste van de bemonsterde locaties binnen het natrium en kalium type, voor enkele locaties kon geen dominante soort water worden bepaald. In termen van anionen, hebben een aantal locaties het chloride soort grondwater, één locatie heeft bicarbonaat grondwater en de andere locaties hebben geen dominante anionen. De totale hardheid varieerde tussen 10 en 662 mg/l, 7 monsters vallen binnen de klasse "zacht water", 3 monsters vallen binnen de "matig hard" klasse, 7 monsters vallen onder "hard water" en 3 monsters vallen binnen de "zeer hard" klasse. De berekening van het Na^+ percentage, het residuële natriumcarbonaat (RSC) en de natrium adsorptieverhouding (SAR) gaf aan dat Nyagatare grondwater geschikt is voor irrigatie. Het Nyagatare district heeft overvloedige graniet en granieten stollingsgesteenten, en dit kan de bron van het fluoride in het grondwater zijn. De bron van de elektrische geleidbaarheid, totale opgeloste stof (TDS), ammoniak en nitriet in het Nyagatare grondwater kan worden gerelateerd aan menselijke activiteiten, zoals bijv. toepassing van kunstmest en dierlijke mest.

Principale componenten analyse (PCA) resultaten toonden aan dat de berekende componenten de variabelen goed weergeven. De zes onderscheidene componenten dragen voor bijna 94% bij tot de variabiliteit in de oorspronkelijk 22 variabelen, zodat men door het gebruik van deze 6-componenten set de complexiteit van de gegevens aanzienlijk kan verminderen, met slechts 6% verlies aan informatie. De eerste component is het hoogst gecorreleerd met fluoride, sulfaat en pH; de tweede component is het sterkst gecorreleerd met calcium en de totale hardheid, terwijl de derde component het sterkst correleert met de totale alkaliteit. De vierde, vijfde en zesde component zijn het meest gecorreleerd met, respectievelijk, kalium, ijzer en magnesium.

De effecten van calcium op de evenwicht adsorptiecapaciteit van As(III) en As(V) aan ijzeroxide gecoat zand en granulair ijzerhydroxide werd onderzocht door middel van batch experimenten, snelle kleinschalige kolomtesten en kinetische modellering. Batch experimenten toonden aan dat bij calcium concentraties $\leq$ 20 mg/l, een hoog As(III) en As(V) verwijderingrendement wordt behaald met zowel IOCS en GFH bij pH 6. Een verhoging van de calcium concentratie tot 40 en 80 mg/l keerde deze trend om, en een hoger verwijderingrendement werd behaald bij hogere pH (8). De adsorptiecapaciteit van IOCS en GFH bij een evenwicht arseen concentratie van 10 µg/l bleken tussen de 2.0 en 3.1 mg/g voor synthetisch water zonder calcium en 2.8 tot 5.3 mg/g bij 80 mg/l calcium te liggen bij alle onderzochte pH-waarden. Na een filtratie van 10 uur in de snelle kleinschalige kolomtesten, wat overeenkomt met ongeveer 1000 Empty Bed Volumes, lagen de C/C_o verhoudingen voor As(V) bij 26% en 18% voor calcium-vrij model water, en slechts 1% en 0.2% na toevoeging van 80 mg/l Ca voor de kolommen gepakt met, respectievelijk, IOCS en GFH. De adsorptie van As(III) en As(V) aan GFH volgt een tweede orde reactie, ongeacht de aanwezigheid van calcium in het model water, terwijl de adsorptie van As(III) en As(V) aan IOCS een eerste-orde reactie volgt in calcium-vrij model water, maar naar een tweede reactie orde kinetiek verschuift in de aanwezigheid van calcium. Het intraparticle diffusiemodel gaf aan dat intraparticle diffusie het belangrijkste controle mechanisme is voor As(III) adsorptie, terwijl de oppervlaktediffusie aanzienlijk bijdraagt bij tot de adsorptie van As(V).

Het effect van PO_4^{3-} op de adsorptieve verwijdering van Cr(VI) and Cd^{2+} werd bestudeerd met IOCS en GFH als adsorbentia. Batch adsorptie experimenten en RSSCT werden uitgevoerd met Cr(VI) and Cd^{2+} houdend model water bij pH 6, 7 en 8.5. De beste Cr(VI) and Cd^{2+} adsorptie werd vastgesteld bij pH 6. GFH toonde een veel betere verwijdering van Cr(VI) dan IOCS, terwijl IOCS beter Cd^{2+} verwijderde in vergelijking met GFH. Toenemende PO_4^{3-} concentraties in het model water van 0 tot 2 mg/L, bij pH 6, induceerde een sterke afname van de Cr(VI) verwijderingefficiëntie, van 93% tot 24% met GFH, en van 24% tot 17% met IOCS. Een soortgelijke trend werd waargenomen bij pH 7 en 8.5. Een uitzondering was de Cr(VI) verwijdering door IOCS bij pH 8.5, welke niet werd beïnvloed door de PO_4^{3-} toevoeging. Cd^{2+} werd goed verwijderd door zowel GFH en IOCS, in tegenstelling tot Cr(VI) dat beter door IOCS werd verwijderd. Het effect van PO_4^{3-} was duidelijk zichtbaar bij pH 6, als er geen neerslag van Cd^{2+} in de oplossing is. Bij pH 8.5 is neerslagvorming het belangrijkste verwijderingproces, namelijk ongeveer 70% van de Cd^{2+} verwijdering. De isotherm constante K voor verschillende combinaties bevestigt de remming van Cr(VI) en de verbetering van Cd^{2+} adsorptie door de toevoeging van PO_4^{3-}. Dezelfde conclusie wordt bevestigd door de resultaten van de snelle kleinschalige kolomproeven. Het mechanisme van de Cr(VI) adsorptie aan GFH en IOCS is waarschijnlijk een combinatie van elektrostatische aantrekking en liganduitwisseling, terwijl het mechanisme van Cd^{2+} verwijdering bij lagere pH (6) was voldoende energierijk om de elektrostatische afstoting te overwinnen.

Ook de effecten van de pH en Ca^{2+} op de adsorptieve verwijdering van Cu^{2+} en Cd^{2+} werden onderzocht in batch adsorptie experimenten en met behulp van kinetisch modelleren. Er werd vastgesteld dat Cu^{2+} and Cd^{2+} niet stabiel waren bij de onderzochte pH-waarden (6, 7 en 8), en dat precipitatie overheerste bij de hogere pH-waarden, vooral voor Cu^{2+}. De toename van de Ca^{2+} concentratie verhoogde ook de precipitatie van Cu^{2+} en Cd^{2+}. Ook is opgemerkt dat Ca^{2+} concurreert met Cu^{2+} en Cd^{2+} voor de bindingsplaatsen aan het oppervlak van het adsorptiemiddel. De aanwezigheid van calcium vermindert het aantal beschikbare adsorptieplaatsen van IOCS en GHF, resulterend in lagere verwijderingefficiëntie voor cadmium en koper. Uit de Freundlich isothermen van de cadmium verwijdering door IOCS bleek dat de adsorptiecapaciteit van IOCS verlaagd als calcium aan het model water werd toegevoegd. De kinetische modellering toonde aan dat de adsorptie van Cd^{2+} aan IOCS waarschijnlijk volgens een tweede-orde reactie verloopt.

De effecten van fulvozuur op de adsorptieve verwijdering van Cr(VI) en As(V) werd ook onderzocht. Batch adsorptie experimenten en karakterisering van IOCS en GFH door middel van SEM/EDS werden gedaan bij verschillende pH-waarden (6, 7 en 8). De atomaire samenstelling van het oppervlak van vers IOCS bleek voor ongeveer 60 tot 75% te bestaan uit, respectievelijk, Fe en O, terwijl koolstof voor ongeveer 10%. De oppervlakte analyse van GFH toonde aan dat Fe en O ongeveer 32% en 28%, respectievelijk, van de chemische samenstelling vertegenwoordigen. Uit de adsorptie experimenten waarin enerzijds As(V) en FA simultaan aanwezig zijn, en anderzijds Cr(VI) samen met FA, blijkt dat de rol van FA onbeduidend is bij bijna alle onderzochte pH waarden, voor zowel IOCS als GFH. Enige interferentie van FA op de verwijdering van As(V) door IOCS of GFH werd alleen waargenomen bij pH 6. Er werd ook vastgesteld dat organisch materiaal (OM) uitloogt uit het IOCS tijdens de experimenten. Het gebruik van EEM toonde aan dat humus-achtige, fulvozuur-achtige en eiwit-achtige organische stof fracties aanwezig zijn in de IOCS structuur.

Verwijdering van bepaalde zware metalen, namelijk Cd^{2+}, Cu^{2+} en Pb^{2+} door IOCS werd ook onderzocht in een reeks batch adsorptie experimenten bij verschillende pH. De bestudeerde metalen waren aanwezig in model water, alleen of samen met enkele andere metalen. De resultaten van de adsorptie-experimenten met model water met één enkel metaal en met IOCS als adsorbens toonden aan dat alle onderzochte metalen zeer effectief kunnen worden verwijderd, met een totaal verwijderingrendement hoger dan 90% bij alle onderzochte pH-waarden. XRF analyse toonde aan dat IOCS voornamelijk hematiet (Fe_2O_3) bevat (ongeveer 85% van de totale massa van mineralen die door XRF kunnen worden geïdentificeerd). Chemische analyse onthulde dat het hoofdbestanddeel van IOCS ijzer is, met name 32% op massabasis. Potentiometrische massa titratie (PMT) gaf een waarde voor de pH met nulpunt lading van 7.0. Het percentage metalen verwijderd door precipitatie bleek metaalspecifiek: het hoogst voor Cu (25%) en het laagst voor Cd (2%) bij pH 8. De gelijktijdige aanwezigheid van concurrerende metalen had geen uitgesproken effect op de totale metalen verwijderingefficiëntie, waarbij de waargenomen vermindering van de totale verwijderingefficiëntie van Cu, Cd en Pb slechts tussen de 1% en 4% lag. Wat de adsorptiecapaciteit betreft werd geen competitief effect van de metalen waargenomen, behalve bij Pb en Cu bij pH 8, waarbij de adsorptie met, respectievelijk, 13% en 22% verlaagd werd.

Complexatie modellering vertoonde twee soorten complexen, een type geassocieerd met een zwakke adsorptieplaats (Hfo_wOCd$^+$, Hfo_wOCu$^+$, Hfo_wOPb$^+$), en een ander type geassocieerd met een sterke adsorptieplaats (Hfo_sOCd$^{+,}$ Hfo_sOCu$^+$, Hfo_sOPb$^+$), die gevormd werden bij alle onderzochte metalen. Precipitatie van Pb en Cu, waargenomen in batch experimenten, werd bevestigd door te modelleren bij pH $\geq$ 6.75. IOCS, dat een goedkoop en gemakkelijk beschikbaar adsorbens is, kan worden gebruikt om water verontreinigd met zware metalen zoals Cd, Cu en Pb te behandelen. Echter, de pH is een belangrijke factor om rekening mee te houden indien men het neerslaan van metalen wil voorkomen die het vloeibaar afval (spoelwater) terecht zullen komen, vooral bij de verwijdering van Cu en Pb.

Publications

Journals

V. Uwamariya, B. Petrusevski, Y.M. Slokar, C. Aubry, P. N.L. Lens and G. Amy (2013), Effect of Fulvic Acid on Adsorptive Removal of Cr(VI) and As(V) from Groundwater by Iron Oxide Based Adsorbents, *Journal of Water, Air and Soil Pollution*, submitted.

V. Uwamariya, B. Petrusevski, P. N.L. Lens and G. Amy (2013), Effect of Phosphate on Adsorptive Removal of Chromium and Cadmium from Groundwater by Iron Oxide Based Adsorbents, *Journal of Environmental Science and Technology*, submitted

V. Uwamariya, B. Petrusevski, P. N.L. Lens and G. Amy (2013), Effect of Calcium on Adsorptive Removal of Arsenic from Groundwater by Iron Oxide Based Adsorbents, *Journal of Environmental Technology*, submitted

V. Uwamariya, B. Petrusevski, P. N.L. Lens and G. Amy (2013), Effect of Calcium on Adsorptive Removal of Copper and Cadmium from Groundwater by Iron Oxide Based Adsorbents, *Journal of Water Supply: Research and Technology-AQUA*, submitted

Conferences

V. Uwamariya, B. Petrusevski, P. Lens, G. Amy (2011), *Effect of Phosphate on Chromium Removal from Groundwater by Iron Oxide based adsorbents, in proceeding of the* IWA Specialist Groundwater conference, *Belgrade 8-10 September 2011.*

V. Uwamariya, B. Petrusevski, P. Lens, N.S. Slokar, N. Stanič, G.Amy *(2011), Removal of Heavy Metals (from groundwater) by Iron Oxide Coated Sand, In proceeding of the* 2nd IWA Development Congress and Exhibition, *Kuala Lumpur (Malaysia), November 21-24, 2011.*

N. Stanič, V. Uwamariya, B. Petrusevski, N.S. Slokar, , G. Amy (2011)*, Competitive Removal of Selected Heavy Metals Commonly found in Urban Storm water Runoff,* the *2nd IWA Development Congress and Exhibition,* Kuala Lumpur (Malaysia), November 21-24, 2011.

V. Uwamariya, B. Petrusevski, P. Lens, G. Amy (2012), Effect of water matrix on adsorptive removal of heavy metals from groundwater. In proceeding of the *IWA World Water Congress and Exhibition, Busan (Korea)* 16-21 September 2012.

Curriculum vitae

Mrs. Valentine Uwamariya was born at Shangi Sector in Nyamasheke District on the 14th May 1971. She went to Groupe scolaire Sainte Famille Nyamasheke for secondary education and later joined the National University of Rwanda. She graduated with the degree of Bsc in Organic Chemistry in 1999 with distinction. From June 2000, she has been working as assistant lecturer at the same university, in the Chemistry Department. In 2003, she joined the University of Witwatersrand in South Africa where she obtained a Master of Science (Electrochemistry) in 2005 with distinction. In the same year she has been promoted to the grade of lecturer. In 2007, she was awarded a scholarship by the Netherlands Government to study a PhD at UNESCO-IHE, Institute for water Education, under sandwich construction program.

Her address in Rwanda is:
National University of Rwanda, P.O. Box 56 Butare, Rwanda
Tel. +250 783 573 023
E-mail: valuwamariya@nur.ac.rw or vuwamariyava@gmail.com

T - #0428 - 101024 - C16 - 240/170/9 - PB - 9781138020641 - Gloss Lamination